Genetic Engineering in Microbial Biotechnology

Genetic Engineering in Microbial Biotechnology

Guest Editor

Fernando Santos-Beneit

MDPI

Basel • Beijing • Wuhan • Barcelona • Belgrade • Novi Sad • Cluj • Manchester

Guest Editor
Fernando Santos-Beneit
Department of Microbiology
and Genetics
University of Salamanca
Salamanca
Spain

Editorial Office
MDPI AG
Grosspeteranlage 5
4052 Basel, Switzerland

This is a reprint of the Special Issue, published open access by the journal *International Journal of Molecular Sciences* (ISSN 1422-0067), freely accessible at: https://www.mdpi.com/journal/ijms/special_issues/HI5063N40H.

For citation purposes, cite each article independently as indicated on the article page online and as indicated below:

Lastname, A.A.; Lastname, B.B. Article Title. *Journal Name* **Year**, *Volume Number*, Page Range.

ISBN 978-3-7258-5345-8 (Hbk)
ISBN 978-3-7258-5346-5 (PDF)
https://doi.org/10.3390/books978-3-7258-5346-5

Contents

About the Editor

Fernando Santos-Beneit

Fernando Santos-Beneit is a microbiologist affiliated with the University of Salamanca, where his key foci are microbial biotechnology and genetic engineering. His expertise spans genetic regulation, antibiotic discovery, advanced microbial chassis engineering, and plastic upcycling, through which he contributes to the development of sustainable circular bioeconomy strategies and the discovering of new biomolecules.

Santos-Beneit obtained his PhD in Biology from the University of León, receiving the Extraordinary Doctorate Award, and has held postdoctoral and research positions in institutions including the University of Valladolid, the University of Oviedo (as a Marie Sklodowska-Curie Fellow), the Center for Bacterial Cell Biology of Newcastle University, the Institute of Biomedicine and Biotechnology of Cantabria, the Vrije University of Amsterdam, and the Institute of Biotechnology of León.

In the course of his career, he has authored over fifty scientific publications in high-impact journals, including *Nature Communications*, and has participated in and led both national and European Union-funded projects. He holds an international patent in the area of plastic upcycling and supervises doctoral theses while developing a research line in creating and optimizing microbial chassis. His work includes studying bacterial mechanisms of antibiotic production, division, and regulation, discovering novel antibiotic resistance mechanisms, and developing microbial strategies for the degradation of plastics. Through these accomplishments, Fernando Santos-Beneit has established himself as a prominent researcher bridging fundamental molecular microbiology with applications aimed at global health and environmental sustainability.

Preface

For millennia, microorganisms have quietly shaped human civilization, enabling the production of bread, beer, and wine. Today, microbial biotechnology stands as a cornerstone of modern science and industry, offering invaluable bioactive compounds such as antibiotics, antifungals, and anticancer agents, while also providing the enzymatic tools and genetic elements that made recombinant DNA technology possible.

This Reprint is devoted to exploring the dynamic interplay between microbial biotechnology and genetic engineering—two disciplines so deeply interconnected that they continuously reinforce each other. While microorganisms fuel advances in genetic engineering by providing enzymes, vectors, and model systems, genetic engineering empowers researchers to reprogram microbial systems with unprecedented precision. From CRISPR-Cas genome editing tools to the engineering of biopolymers and the enhancement of antibiotic production, the synergy between these fields continues to expand the frontiers of medicine, industry, agriculture, and environmental sustainability.

The scope of this Reprint extends across the diverse colors of biotechnology: red, white, yellow, and green. Aiming to showcase current achievements and inspire further exploration of how the smallest organisms, microorganisms, can address the largest challenges, such as climate change, pollution, and global health, this Reprint is addressed to researchers, clinicians, engineers, and students who seek to understand not only the mechanisms underlying microbial innovation but also its transformative potential for society.

Our motivation for assembling this collection is both scientific and humanistic: to highlight rigorous research, foster interdisciplinary collaboration, and encourage a vision in which microbes and molecular tools contribute to a healthier and more sustainable future.

Fernando Santos-Beneit
Guest Editor

International Journal of
Molecular Sciences

MDPI

Review

Genetic Modifications in Bacteria for the Degradation of Synthetic Polymers: A Review

Diego Martín-González [1,†], Carlos de la Fuente Tagarro [1,†], Andrea De Lucas [1], Sergio Bordel [1,2] and Fernando Santos-Beneit [1,2,*]

1 Department of Chemical Engineering and Environmental Technology, School of Industrial Engineering, University of Valladolid, Dr. Mergelina, s/n, 47011 Valladolid, Spain; diego.marting@uva.es (D.M.-G.); andrea.lucas@uva.es (A.D.L.); sergio.bordel@uva.es (S.B.)
2 Institute of Sustainable Processes, Dr. Mergelina s/n, 47011 Valladolid, Spain
* Correspondence: fernando.santos.beneit@uva.es
† These authors contributed equally to this work.

Abstract: Synthetic polymers, commonly known as plastics, are currently present in all aspects of our lives. Although they are useful, they present the problem of what to do with them after their lifespan. There are currently mechanical and chemical methods to treat plastics, but these are methods that, among other disadvantages, can be expensive in terms of energy or produce polluting gases. A more environmentally friendly alternative is recycling, although this practice is not widespread. Based on the practice of the so-called circular economy, many studies are focused on the biodegradation of these polymers by enzymes. Using enzymes is a harmless method that can also generate substances with high added value. Novel and enhanced plastic-degrading enzymes have been obtained by modifying the amino acid sequence of existing ones, especially on their active site, using a wide variety of genetic approaches. Currently, many studies focus on the common aim of achieving strains with greater hydrolytic activity toward a different range of plastic polymers. Although in most cases the depolymerization rate is improved, more research is required to develop effective biodegradation strategies for plastic recycling or upcycling. This review focuses on a compilation and discussion of the most important research outcomes carried out on microbial biotechnology to degrade and recycle plastics.

Keywords: synthetic polymers; plastics; biodegradation; genetic engineering; PET; PETase; cutinase; esterase

1. Introduction

1.1. Definition and Classification of Plastics

Plastics are a type of synthetic polymeric material based on carbon and hydrogen and with high molecular weight [1]. Industrial-scale plastic production began in the 20th Century, and their use has increased over time [2,3]. Plastics can be classified according to different criteria. For example, depending on the composition of their backbone, they can be homochain polymers (when their backbone is made up exclusively of carbon) or heterochain polymers (if there are other elements, like oxygen and nitrogen, present in their backbone) [4]. Depending on the monomers that they are made of, they can be classified as homopolymers if they only have one monomer or copolymers if they have two or more different monomers [1]. Depending on their thermomechanical properties, they can be thermosets if, during their fabrication, covalent bonds are established between polymer chains, giving them high resistance and a shape that cannot be modified thermically; thermoplastics, if no covalent bonds are established between polymer chains so they can be fused and reshaped; or elastomers if they have a high elasticity [5]. Depending on the presence of benzene rings in their backbone, they can be aromatic if they have benzene rings or aliphatic if they do not have benzene rings [6]. Depending on the raw material they are made from, they can be petrochemical (the raw material is petroleum) or biobased (the raw material is biomass), and depending on their degradability by organisms, they can be biodegradable or non-biodegradable.

1.2. Advantages and Disadvantages of Plastics

In general, plastics are lightweight, long-lasting, inert and cheap and easy to produce. Their diversity is such that there are plastics with the ideal characteristics for almost any application. Furthermore, the use of mixtures and alloys, as well as additives, can change the properties of the material and adjust them as desired [1]. As plastics are remarkably diverse and can have vastly different properties, their uses are just as varied. They are used to make, among many others, fibers and textiles, toys, packaging, healthcare instruments, such as syringes and implants, and construction materials for insulation, pipes and cable coatings [1]. Even though plastics are generally cheap and easy to produce and have good mechanical properties, the field of engineering requires especially resistant materials for very specific applications. Though they are more expensive, the so-called "engineering plastics" are used for that purpose as they boast a high performance [1]. Other plastics, known as commodity plastics, are found in everyday items.

The most used commodity plastics are poly(ethylene terephthalate) (PET), high-density polyethylene (HDPE), poly(vinylchloride) (PVC), low-density polyethylene (LDPE), polypropylene (PP) and polystyrene (PS) [2,7,8]. Products made with these plastics can be identified according to the ASTM International Resin Identification Coding System (RIC). The "Others" category includes all other plastics, as well as products made with a mixture of two or more plastics [9]. The global annual production of plastics exceeded 400 million metric tonnes (400 Mt) in 2022, 362.3 of which were new petrochemical plastics. Commodity plastics make up the following percentages (see Figure 1): PET 6.2%, HDPE 12.2%, PVC 12.7%, LDPE 14.1%, PP 18.9% and PS 5.2% [10]. Out of the different uses, most plastic is destined for packaging. In 2019, 142 Mt (31%) of plastics were used in packaging [3,7]. Between 1950 and 2022, more than 10 billion metric tonnes (10.000 Mt) of plastics were produced [2,10,11].

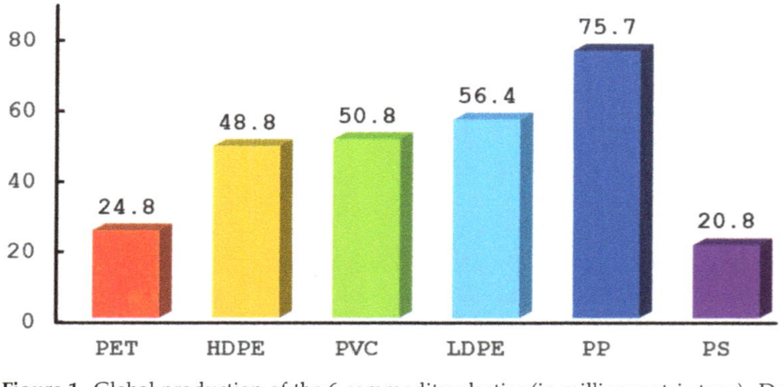

Figure 1. Global production of the 6 commodity plastics (in million metric tons). Data obtained from [10].

Despite their useful properties, the use of plastics has two main disadvantages, which have to do with the beginning and the end of their lifetime: the first one is the fact that most plastics are petrochemical, and the second one is the amount of non-biodegradable waste generated. Regarding their origin, the reserves of petroleum are limited as petroleum is a non-renewable resource, and given the production rate in 2020, it is estimated that reserves will last for 50 years [12]. For the purposes of this article, the waste generated is the most relevant problem.

2. Bioplastics as an Alternative to Petroleum-Based and Non-Biodegradable Plastics

The most recent assessments estimate that out of the 9200 Mt of plastics produced until 2017, 6500 Mt (70%) have become waste. Of this waste, 5000 Mt (54%) ended up in landfills or were released into the environment [13]. This plastic waste has reached virtually every corner of the planet. The five so-called garbage patches are notorious for accumulating 250,000 tons of plastic [7], but plastics are also found elsewhere. They have been found in rivers, lakes, drinking water and table salt and even in extremely remote places, such as the seabed and both polar regions. They have reached such areas mostly in the form of micro- and nanoplastics, whose size allows them to travel much further and even enter organisms' cells [14]. Due to their composition and structure, plastics can remain in the environment for thousands of years [15], and their effects on living beings' health, while a complicated matter that is still under study, are undoubtedly severely negative. This has been researched especially in marine ecosystems, where they can cause harm to many different organisms [8,16,17], but they are also apparent in terrestrial ecosystems [18] and human health [19,20]. Moreover, the additives used to modify the properties of plastics can also have toxic effects on organisms [21], and plastics can adsorb and accumulate heavy metals, persistent organic contaminants and even pathogens [22].

Due to their harmful effects on the environment, plastics, like any other type of waste, have to be managed appropriately. Recycling rates are overall very low, with only 600 Mt (6.5%) of plastics produced worldwide until 2017 having been recycled and more than 5000 Mt of them ending up in landfills or being released into the environment [13]. Landfilling plastic waste, though simple, is not an adequate waste management strategy, as landfills can leach microplastics [23,24] as well as other contaminants [25]. Moreover, plastic waste in the environment does not remain permanently unchanged, as it slowly undergoes physicochemical changes that partially break it down. These changes include photodegradation, hydrolysis and thermo-oxidation. However, this transformation results mostly in the fragmentation of plastic rather than its mineralization, generating smaller plastic fragments (micro- and nanoplastics) that still remain in the environment. In order for plastics to be mineralized and, therefore, completely removed from the environment, biological degradation is required [26]. This biological degradation can be performed by different organisms but, even if plastics end up mineralized by the action of these organisms, the degradation rates are so slow and the plastic permanence is so high that most plastics are considered non-biodegradable. Given the problems derived from the properties of the most used plastics, a new type of material has received a lot of attention: bioplastics.

Bioplastics are a type of plastic that differs from conventional plastic in at least one of two ways: the raw material they are made from or their biodegradability. Bioplastics can either be made from biomass—biobased plastics—or be biodegradable or have both properties [27]. A particular type of bioplastics is poly(hydroxyalkanoates) (PHAs). They are naturally produced by some microorganisms, mostly bacteria, for energy storage [28] and are both biobased (as they are produced by an organism that can be grown with renewable feedstock) [29] and biodegradable (as they can be easily enzymatically degraded) [30]. Just like plastics in general, bioplastics are very varied, so grouping them in the bioplastic category is not particularly informative of their properties, especially without specifying which criteria make them such. Therefore, when discussing individual plastics, this work will specify if they are petrochemical or biobased and biodegradable or non-biodegradable (Figure 2). The main benefits of using bioplastics are summarized in the fact that biobased plastics do not involve the use of petroleum, and biodegradable plastics can be fully mineralized by organisms. However, bioplastic production still has an environmental impact when the whole life cycle is considered [31], and regarding biodegradable plastics, this property does not mean that they can be released into the environment without consequences [32].

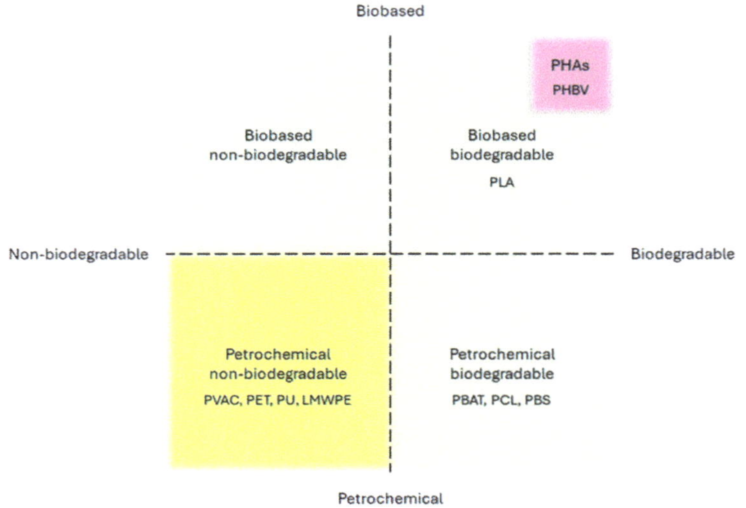

Figure 2. Classification of plastics according to the raw material and biodegradability. PHAs (including a major type, i.e., PHBV) constitute a special category of plastics since they are completely natural.

Biodegradable plastics can remain in the environment for long periods of time [33–36] and carry potentially toxic substances [37]. They can even act as a reservoir of microbes with antibiotic-resistance genes [38], so their waste should be managed, too [39]. In the case of biodegradable plastics, the preferred waste management strategy is carrying out their degradation by organisms. There are international standards developed to clearly define the conditions and timescale of biodegradation [40–46], but their use is not mandatory, so the criteria to determine the biodegradability of plastics can be quite arbitrary. However, plastic is generally considered biodegradable if there are organisms that can mineralize it in a reasonable amount of time in controlled conditions, which must always be specified.

3. Biodegradation of Plastic Polymers

The last few decades have seen an increasing interest in both biodegradable plastics and organisms that can degrade plastics. As new species and strains are discovered and genetic engineering is used to improve the enzymes involved in biodegradation, plastics that were previously considered non-biodegradable can now be degraded by one organism or another and to a greater or lesser extent. Therefore, the categories of biodegradable and non-biodegradable plastics are not fixed and can change over time, but in this work, we will use the usual classification.

Plastic-degrading microorganisms include bacteria, fungi and microalgae [47]. This review focuses on bacterial strains. Although genetic engineering approaches have been performed for decades, only a few bacteria have been modified to improve their capability of degrading these synthetic plastics or to grant them this characteristic by modifying their metabolism. More specifically, the modifications performed consist of the heterologous expression of enzymes that come from other organisms, mutating/changing specific amino acids of the enzymes and the addition of domains or other protein structures to the enzymes. Different techniques have been used in order to carry out these modifications, such as metagenomic analysis, directed and non-directed mutagenesis and the use of fusion proteins or chimeras.

This review is a compilation of the studies in which bacteria have been genetically modified, whether to grant them or enhance their ability to degrade plastics or simply to perform genetic modifications and will only mention the polymers for which modified bacteria have been obtained. Table 1 shows the modified bacterial strains that have been developed for the degradation of plastic polymers according to research published in the scientific literature (i.e., PubMed, Scopus, etc.).

Table 1. Heterologous expression of enzymes to degrade PET, PLA, PBAT, PVAC, PCL, PBS, PHV, PU and LMWPE.

Host	Plastic	Enzyme	Origin Species	Ref.
E. coli BL21 (DE3)	PLA	Protease (Plasmid: pET26b(+))	*Thermus* sp. Rt41A	[48]
E. coli BL21-Gold (DE3)	PBAT	Esterases Cbotu_EstA and Cbotu_EstB	*Clostridium botulinum*	[49]
E. coli DH5α	PVAC, PCL	Cutinase (Cut) and lipase (Lip) (Plasmid: pPICZαA)	*Thermomyces lanuginosus* (Lip); *Thielavia terrestris* NRRL 8126 (Cut)	[50]
E. coli XL-10	PET, PBS, PHBV	Cutinase 1 (Thc_Cut1) (Plasmid: pMK-T; pPICZαB)	*Thermofida cellulosilytica*	[51]
E. coli XL10-Gold	PU	Polyamidase (PA) (Plasmid: pET26b(+))	*Nocardia farcinica* IMA 10152A (PA)	[52]
E. coli BL21 Gold (DE3)	PU	Polyamidase (PA) (Plasmid: pET26b(+))	*Nocardia farcinica* IMA 10152A (PA)	[52]
E. coli BL21	LMWPE	Alkane hydroxylase (Plasmid: pUC19)	*Pseudomonas* sp. E4	[53]
E. coli BL21-CodonPlus (DE3)	PET, PCL	Cutinase	Metagenomical library	[54]

Since PET is one of the most produced plastics worldwide and its degradation is challenging, a special focus is put in this review in relation to the microbial biodegradation strategies developed specifically for this polymer (summarized in Table 2).

Table 2. Heterologous expression of enzymes to degrade PET.

Host	Plastic	Enzyme	Origin Species	Ref.
E. coli BL21 (DE3)	PET	Cutinase (Plasmid: pET25b(+))	*Fusarium solani pisi*	[55]
E. coli BL21 (DE3)	PET	Cutinase Tfu_0883 (Plasmid: pET20b)	*Thermobifida fusca*	[56]
E. coli BL21-Gold (DE3)	PET	Cutinase Thc_Cut2 (Plasmid: pET26b(+))	*Thermobifida cellulosilytica* DSM44535	[57]
E. coli DH5α	PET	Cutinase-type polyesterase (Cut190) (Plasmid: pGEM-T; pQE80L)	*Saccharomonospora viridis* AHK190	[58]
E. coli Rosetta-gami B (DE3)	PET	Cutinase-type polyesterase (Cut190) (Plasmid: pGEM-T; pQE80L)	*Saccharomonospora viridis* AHK190	[58]
E. coli BL21 (DE3)	PET	Cutinase TfCut2	*Thermobifida fusca* KW3	[59]
E. coli BL21 (DE3)	PET	Cutinase TfCut2, LC-Cutinase, carboxyl esterase TfCa (Plasmid: pET-20b(+))	*Thermobifida fusca* KW3	[60]
E. coli XL1-Blue	PET	PETase (Plasmid: pET32a)	*Ideonella sakaiensis* 201-F6	[61]
E. coli BL21 DE3	PET	LC-cutinase (Plasmid: PET28; PJ912)	Plant compost metagenome	[62]

Table 2. *Cont.*

Host	Plastic	Enzyme	Origin Species	Ref.
E. coli Rosetta-gami B	PET	PETase (Plasmid: pET15b; pET15a)	*Ideonella sakaiensis*	[63]
E. coli BL21-CodonPlus (DE3) RIPL	PET	PETase (Plasmid: pET-21b)	*Ideonella sakaiensis*	[64]
E. coli C41 (DE3)	PET	PETase (Plasmid: pET-21b(+))	*Ideonella sakaiensis* 201-F6	[65]
E. coli BL21 (DE3)	PET	PETase (Plasmid: pET28a)	*Ideonella sakaiensis*	[66]
E. coli Rosetta-gami B	PET	Multiple modified *Is*PETase (*Is*PETaseS121E/D186H/R280A)	*Ideonella sakaiensis*	[67]
E. coli DH5α	PET	LC cutinase (Plasmid: pHK-LCC)	Plant compost metagenome	[68]
E. coli BL21 (DE3)	PET	LC cutinase (Plasmid: pHK-LCC)	Plant compost metagenome	[68]
Clostridium thermocellum DSM1313	PET	LC cutinase (Plasmid: pHK-LCC)	Plant compost metagenome	[68]
E. coli BL21 (DE3)	PET	Hydrolases 1 and 2 (BTA1 and BTA2); cutinase (FsC); *Is*PETase; leaf-branch compost cutinase (LCC)	*Thermobifida fusca* (BTA1 and BTA2); *Fusarium solani pisi* (FsC); *Ideonella sakaiensis* 201-F6 (*Is*PETase); leaf compost metagenome (LCC)	[69]
E. coli MG1655 RARE	PET	Terephthalate 1,2-dioxygenase, dihydroxy-3,5-cyclohexadiene-1,4-dicarboxylic acid dehydrogenase, carboxylic acid reductase, catechol O-methyltransferase	*Ideonella sakaiensis*	[70]
E. coli PHL628	PET	PETase bound to BIND platform	*Ideonella sakaiensis*	[71]
E. coli TOP10	PET	PETase bound to BIND platform	*Ideonella sakaiensis*	[71]
E. coli BL21 (DE3)	PET	Leaf-branch compost, cutinase (LCC) and variants	Leaf compost metagenome	[72]
E. coli DH5α	PET	Leaf-branch compost, cutinase (LCC) and variants	Leaf compost metagenome	[72]
E. coli NEB5α	PET	*Is*PETase; *Is*PETase-MHETase chimera	*Ideonella sakaiensis*	[73]
Vibrio natriegens	PET	*Is*PETase; *Is*PETase-MHETase chimera	*Ideonella sakaiensis*	[73]

3.1. Biodegradation of PET

PET is one of the most produced petrochemical synthetic polymers, accounting for around 6.2% of global plastic production in 2022 [10], and its market size surpasses other highly produced plastics, such as HDPE, PVC or PP [74]. The main reason why PET is so produced is because its molecular structure offers a lot of versatility, making it essential in our daily lives. PET is a long semi-aromatic thermoplastic polyester chain produced from ethylene glycol (EG) and terephthalic acid (TPA). Its production has two steps. First, the union of two EG molecules and one TPA molecule by esterification generates an intermediate molecule, bis(2-hydroxyethyl) terephthalate (BHET). Secondly, BHET along with catalysts, such as Sb_2O_3 or $Sb(OAc)_3$, is subjected to a process of polymerization, creating the long chain through ester bonds (Figure 3A) [5,75]. When it comes to the manufacturing process, amorphous and semi-crystalline PET are produced depending on the thermal processing undergone during the polymerization. The main difference between these two types lies in the intrinsic viscosity and molecular weight. In the case of the amorphous polymer, the long polyester chains are randomly set out, resulting in a more flexible plastic. On the other hand, semi-crystalline materials are formed by amorphous domains and chains arranged in an orderly way, making the material more resistant and less ductile (Figure 3B). These differences in its molecular structure make it both chemically and thermally stable. This is what makes PET a strong and durable compound, ideal for a wide variety of applications, such as synthetic fibers for the textile industry, water bottles and packaging [5,76].

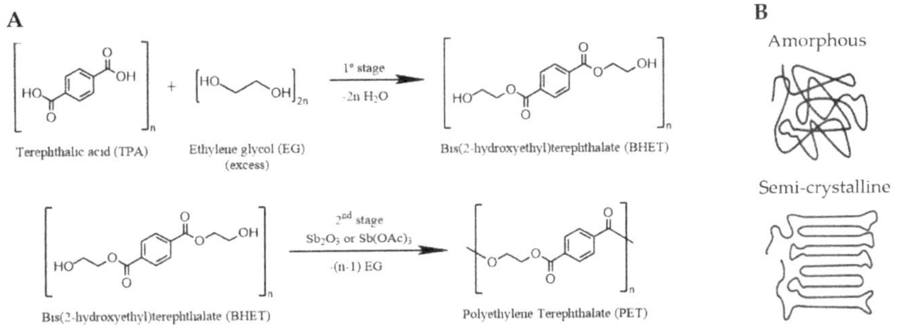

Figure 3. (**A**) PET industrial formation process using BHET as intermediate. Based on [5]. (**B**) Molecular structure of amorphous (**up**) and semi-crystalline (**down**) materials of PET. Adapted from [77].

Despite the advantages provided by PET, it also presents the serious drawback of what to do with it after its useful life. Most PET comes to an end accumulated in landfills. It is estimated that it takes around 300 years to decompose, degrading over time because of solar radiation and heat, among other factors, and releasing harmful chemical compounds to the environment. The estimate increases if its degradation is not accelerated by heat or solar radiation, reaching 2500 years or more. PET is so difficult to degrade due to its physicochemical properties, which make it resistant to decomposition by water and organic and inorganic compounds [73,76].

For this reason, there are a series of methods, which involve mechanical, chemical and biological methods, to recycle and reuse this polymer, producing fibers and fabrics. The mechanical methods are the most widespread, but they are also very expensive, while in the case of chemical methods, the process can be harmful to the environment [70,78].

In contrast with the previously mentioned methods, biological methods are on the rise because they do not damage the environment. These methods are still in development and are based on the use of cells as factories of enzymes able to break the bonds of PET, releasing the monomers of which it is composed. After that, those different monomers can be subjected to a valorization process to generate high value-added molecules, such as polyhydroxyalkanoates, vanillic acid, gallic acid, lycopene, glycolic acid, pyrogallol, catechol and muconic acid, which can be used as flavors, cosmetics, sanitizers and animal feed and in pharmacy, among other uses [78,79].

3.1.1. Enzymes Involved in the PET Degradation Pathway

The reason why PET can be degraded by enzymes is because a lot of them are unspecific regarding their substrates. The current literature offers many names given to these enzymes, such as PET hydrolases, PET esterases, PET cutinases, PET depolymerases or PETases. PET esterases, PET cutinases and PET depolymerases are hydrolytic enzymes that are able to break the ester bonds of biological molecules, like suberin and cutin, and due to their unspecificity, it turns out that they can break those of PET as well. On the other hand, PETases are hydrolytic enzymes more specific to PET [80,81]. Due to the specificity, there are small differences between the different types of enzymes, such as PET cutinases, which can hydrolyze the ester bonds of aliphatic and aromatic molecules, while PETases can only hydrolyze bonds of aromatic molecules [54,65,69]. To simplify the following explanation, we will refer to all of them as PETases unless it is otherwise stated, with the understanding that PETases are enzymes that hydrolyze PET.

The hydrolyzation pathway of PET is summarized in Figure 4. In most cases, PETases are capable by themselves of firstly depolymerizing the PET chain and secondly degrading its monomers—mono(2-hydroxyethyl) terephthalate (MHET) and BHET—to finally form TPA and ethylene glycol. PETases can directly produce MHET and TPA, or In contrast, if BHET is generated, they can degrade it to MHET [5,78]. However, there are some PETases that are not able to carry out both steps by themselves, as is the case with PE-H (*Pseudomonas aestusnigri*) and *Is*PETase (*Ideonella sakaiensis*). Both can only depolymerize PET because they cannot break the ester bond of MHET, which accumulates [82]. In contrast, there are enzymes capable of degrading only MHET (MHETases). *Is*MHETase is an example, being able to degrade MHET to finally produce ethylene glycol and TPA. Although there are enzymes that do not carry out both degradation actions by themselves, they can complement each other. *Is*MHETase and *Is*PETase were found in *I. sakaiensis* 201-F6 when this strain was discovered in a medium whose only carbon source was PET [81]. Recently, a new category of enzymes has been discovered, BHETases, which specifically catalyze the transition from BHET to MHET. Two BHETases have been characterized, *Chry*BHETase from *Chryseobacterium* sp. and *Bs*Est from *Bacillus subtilis* [83].

As a consequence of the cooperation between PETases and MHETases or the action of only PETases, PET is hydrolyzed, and TPA and ethylene glycol are released, and they can be metabolized (Figure 4). TPA presents a biochemical pathway to finally be degraded to succinyl-CoA. Regarding ethylene glycol, it can be metabolized to obtain acetyl-CoA [78]. Both succinyl-CoA and acetyl-CoA are incorporated into the tricarboxylic acid cycle (TCA) to obtain energy for the bacteria and, in certain cases, some products of interest. For example, species *I. sakaiensis* and *Geobacter sulfurreducens* when cocultivated can degrade PET to ethylene glycol and generate electricity [84]. In another study, a *Rhodococcus josii* strain PET metabolized the PET hydrolysate and synthesized lycopene, which is used in medicine for cancer treatments [85]. One example of a modified bacteria is *Escherichia coli* modified to degrade PET and produce vanillin, present in cosmetic and food industries [70]. The cleavage of the ester bond happens inside the enzyme's active site, and the ability of different enzymes to hydrolyze PET comes as a consequence of a nucleophilic similarity produced by three amino acids conserved in PETases, Ser-His-Asp. In contrast, the adjacent sequences to these amino acids are different among enzymes, varying their selectivity to substrates [50,63].

There are two proposed classifications for PETases in the literature. One of them is based on their sequence, while the other one is based on their protein structure. According to the sequence-based classification, PETases can be classified as type I and type II enzymes [5]. Type I enzymes have one C-terminal disulfide bridge, and type II enzymes have two, with one of them being close to the active site. This additional disulfide bridge provides type II enzymes more thermal stability and plays a key role in the hydrolytic action [5,63,86]. Another characteristic that differentiates both types is the amino acids present in the active site. Type I enzymes, such as LCC and Cut190, which will be mentioned later, have His159 and Phe/Tyr238 residues, and type II enzymes present Trp and Ser in the same locations. In addition, type II enzymes are subdivided into type IIa if they have Phe or Tyr residues instead of Ser238 or type IIb if they maintain Ser, as is the case with *Is*PETase. As an exception, type I enzymes may not have the disulfide bridge. This is the case with PET27 and PET30 from *Aequorivita* sp. CIP111184 and *Kaistella jeonii*, respectively. According to the structure-based classification, PETases can have three different structures, with *Is*PETase, *Fusarium oxysporum* cutinase and *Chloroflexus* sp. MS-G cutinase representing Structures 1, 2 and 3, respectively [87]. For example, *Is*PETase presents an α/β-hydrolase fold and a core made up of eight β-strands and six α-helices, as well as a highly polarized surface [65].

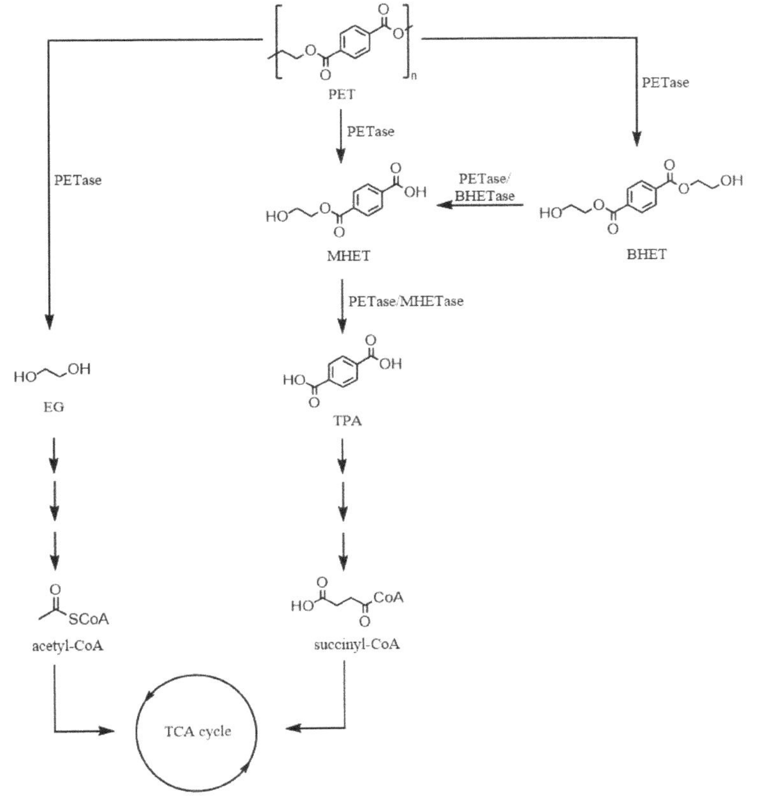

Figure 4. Chemical structures of PET and products related to its hydrolyzation. Each arrow represents an enzymatic reaction. The presence of multiple arrows, like between EG—acetyl-CoA and TPA—succinyl-CoA, refers to the fact that there is more than one chemical reaction between one molecule and another. Adapted from [80].

3.1.2. Modifications of Bacteria and Enzymes to Improve PET Degradation

Since the discovery of PET-degrading enzymes in the last century, efforts have been made to improve their degradation rate, either by searching for new enzymes and heterologously expressing them in other microorganisms or by modifying the existing ones. It was not until 2016 that the first PETase (*Is*PETase) was discovered in *I. sakaiensis* 201-F6, a specific enzyme that allowed the bacteria to use PET as its major carbon and energy source. The experiments were carried out at 30 °C and pH 7, and this enzyme was compared with three other enzymes with hydrolytic activity toward PET [81]. *Is*PETase has an α/β hydrolase fold like cutinases but with a larger active site [65]. After this discovery, modifications of this *Is*PETase have been carried out in order to obtain a higher degradation rate. The importance of the amino acids of the active site or close to it is emphasized, as will be seen afterward with cutinases. In addition, the presence of one amino acid or another may affect the enzyme–substrate interaction affinity and, therefore, the degradation of the polymer [86].

However, before *Is*PETase was discovered, different genetic engineering methods had already been carried out to increase the degradation speed of esterases and cutinases [55]. Site-directed mutagenesis was used to specifically modify the sequence close to the active site of a cutinase from *Fusarium solani pisi*. Mutations were carried out to create a bigger space in the active site to facilitate the entrance of the polymeric chain. The modifications that created a more hydrophobic binding site resulted in the greatest PET degradation compared with the native enzyme. PET was degraded at a rate of 12 mg/h/mg of modified cutinase at pH 8 and 50 °C, while the native enzyme degraded 0.05 mg/h/mg of the enzyme at pH 7.0 and 55 °C.

Continuing with the idea of increasing hydrophobicity, in 2011, two amino acids were removed from the active site of the cutinase Tfu_0883 from *Thermobifida fusca*, expanding the space and improving its capacity to hydrolyze polyesters due to better PET adsorption. In the same experimental conditions, after 4 h, the native enzyme reached a spectral value (k/s) of 0.1, while the modified enzyme reached almost 0.2. Stretching the time, at 48 h, the values were around 0.15 and 0.3, respectively [56]. In addition, the amino acids of surface regions outside the active site turned out to have a direct action in PET hydrolysis, as mentioned in Herrero-Acero et al. [57]. In this study, the cutinase Thc_Cut2 from *Thermobifida cellulosilytica DSM44535* was modified by exchanging amino acids in the surface region. It was shown that exchanging a surface positive residue, like arginine, for a non-charged one, like asparagine and serine, makes a modified enzyme with a higher affinity for PET, according to their k_{cat}/K_m values. In this case, the modifications produced a variation in the hydrolytic capacity of the enzymes, but the hydrophobicity did not change at any time. It is speculated that these changes in the amino acid sequence may decrease the size of the active site, stabilizing the region.

A similar idea is mentioned in Kawai et al. [58], where CUT190 cutinase from *Saccharomonospora viridis* AHK190 was triple-mutated to Ser226Pro/Arg228Ser/Thr262Lys. It was previously known that in its native state, it is able to degrade different polymers, such as PCL, PBSA, PBS and PHB, among others. However, these modifications resulted in an enzyme with more activity and higher thermostability. A possible explanation for this result was that a polar amino acid with a positive charge, such as arginine, close to the binding site, could interact with negative charges that may be present in the substrate molecules [61]. For this reason, arginine was replaced by serine. In addition, it was concluded that the addition of Ca^{2+} to the medium provides greater thermostability to CUT190 [58]. In another study, Han et al. [61] obtained an enzyme with a mutation near the substrate binding site that facilitated the union with the substrate and increased its thermostability to 5 °C higher than the native enzyme.

In contrast, Austin et al. [65] achieved a modified *Is*PETase exchanging two amino acids—Ser238Phe/Trp159His—conserved in cutinases next to the active site. The modified enzyme had a smaller binding site and resulted in a slightly higher degradation rate of PET and polyethylene-2,5-furandicarboxylate (PEF), a semiaromatic polyester derived from PET. Hydrophobicity is mentioned to be relevant in the adsorption of the polymer by the active site. In the following work with *Is*PETase, eight modified enzymes were obtained. Three of them with modifications in their active site—Arg61Ala, Leu88Phe and Ile179Phe—resulted, respectively, in higher PET degradation rates of 13.5 mg PET/μmol PETase·L^{-1} per day, 17.5 mg PET/μmol PETase·L^{-1} per day and 22.5 mg PET/μmol PETase·L^{-1} per day compared with the native *Is*PETase of 8.2 mg PET/μmol PETase·L^{-1} per day. In two of them, the hydrophobicity remained constant, and in the third, it increased. However, the other five modified enzymes obtained lower results than the native one. In three of these, the hydrophobicity increased, and only in one mutant, it decreased [66]. Therefore, according to these studies, not only the hydrophobicity in the active site can influence PET depolymerization, but the type of amino acid that is modified can also affect it.

Focusing on another parameter, in 2019, it was tested to see if a more thermostable *Is*PETase was able to degrade PET at a faster rate. Native *Is*PETase has a Tm of 48.81 °C, while the obtained triple-mutated *Is*PETase—Ser121Glu/Asp186His/Arg280Ala—had a

Tm of 57.62 °C. This mutant released 14 times more monomers at 40 °C than the native enzyme, obtaining an enzymatic activity of 120.9 μM and 8.7 μM, respectively [67].

In addition, there are genetic engineering works based on metagenomics, such as in Sulaiman et al. [54], where they cloned a leaf-branch compost cutinase (LC-cutinase or LCC), whose highest identity was with a *Thermomonospora curvata* lipase with 59.7%, and degradation of PET was achieved under the conditions of 50 °C and pH 8 at a rate of 12 mg/h/mg of the enzyme. This result is much greater when compared to *T. fusca* cutinase at 55 °C and pH 7, which consumes 0.05 mg of PET/h/mg of the enzyme [88].

The main drawback of expressing LC-cutinase is that it forms aggregates with itself due to electrostatic interactions as a result of increasing the temperature. This produces an inhibition in the depolymerization of PET. For this reason, it was tested to see if glycosylating LC-cutinase could prevent this. The result was that glycosylated LC-cutinase began to form aggregates at 10 °C above its native protein [62].

Following the idea that glycosylation can influence enzymes' properties, another study used the previously mentioned cutinase Thc_Cut1 from *Thermobifida cellulosilytica* and generated two mutated cutinases without glycosylation sites. All were able to hydrolyze polyesters, such as PET, poly(butylene succinate) (PBS) and poly(3-hydroxybutyrate-*co*-3-hydroxyvalerate) (PHBV), but one of the mutants had higher PBS-degrading activity than the native enzyme [51]. Metagenomics was also used by Wei et al. [59], exchanging amino acids from the structure of a cutinase obtained with this technique for others from TfCut2, a polyester hydrolase from *Thermobifida fusca KW3*. The best mutants were those with modifications near the binding site. The mutations Gly62Ala and G62Ala/I213Ser achieved a weight loss of 42% of PET films, almost three times more compared to the native protein. This may be due to the fact that the wild-type enzyme is inhibited by MHET, and the substitution of these amino acids may decrease the affinity of the mutants for this molecule. To avoid the problem of inhibition by MHET, Barth et al. [60] used the dual-protein system with TfCut2 and the carboxylesterase TfCa, both from *T. fusca* KW3. With this, TfCut2 degraded PET, and TfCa achieved the same with BHET and MHET. This dual system generated 2.4 times more TPA than using TfCut2 alone. On the other hand, Tournier et al. [69] obtained a modified LC-cutinase with a higher affinity for PET and higher thermostability. This was achieved by exchanging two amino acids, previously tested using in silico methods, with which the new enzyme presented a disulfide bridge. This bridge conferred greater strength in the bond to the substrate, and the other two modifications improved thermostability. In its native state, this enzyme depolymerized 50% of the amorphous PET, while the modified LC-cutinase reached 90% after 10 h. Regarding productivity, it produced 16.7 g of TPA per liter per hour as a consequence of degrading PET. The optimal ratio to obtain these results was 3 mg of mutated enzyme/per g of PET.

Another article based on metagenomics modified the anaerobic bacteria *Clostridium thermocellum* to constitutively express and secrete LC-cutinase in large quantities. After 14 days incubating anaerobically at 60 °C, more than 60% of amorphous PET was degraded. It was 50 mg in total, corresponding to more than 2.2 mg each day [68].

One interesting aspect is obtaining high-value by-products through the degradation of PET, as shown by Sadler et al. [70]. Following this idea, they created the strain *E. coli* MG1655 RARE. This genetically modified strain is able to degrade PET, release TPA and obtain vanillin. This was made using PETases from *I. sakaiensis* and a de novo pathway to convert TPA to vanillin. To optimize the process, parameters such as temperature and cell permeabilization were improved, and 79% vanillin production was obtained compared to TPA.

In recent years, there has been a trend to create fusion proteins to depolymerize PET. This is the case with Zhu et al. [71]. They modified *E. coli* with the BIND-PETase dimer. It is a dimer in which the so-called biofilm-integrated nanofiber display (BIND) binds to the curli of bacteria. Curli are fibers on the outside of the cell that are involved in forming biofilms. Through this method, PETase bound to BIND remains on the surface of the cell and degrades the polymer. It is a system that provides thermostability to PETase and makes it able to degrade 9.1% semi-crystalline PET and microplastics between 2.5 and 50 mg.

Based on the protein fusion strategy, Chen et al. [72] synthesized two modified LC-cutinases. They based them on the modified LC-cutinase created by Tournier et al. [69] called LCC^{ICCG} and previously mentioned, which had higher catalytic activity and thermal stability but little affinity for PET. These two fusion proteins, LCC^{ICCG}-TrCBM and CfCBM-LCC^{ICCG}, were made by adding a substrate-binding domain to LCC^{ICCG} from *Trichoderma reesei* and *Cellulomonas fimi*, respectively. These domains were selected in silico based on the principles that domains from thermophilic organisms, domains that bind to ordered crystalline substrates and domains that are smaller than the rest of the protein are more adequate. It was estimated with the experimental results that the affinity for the polymer increased 1.4 fold in LCC^{ICCG}-TrCBM and 1.3 fold in CfCBM-LCC^{ICCG}, while they degraded 3.7% and 24.2% more PET, respectively.

Another recent project using this technique appears in Li et al. [73]. The researchers modified the non-pathogenic, moderate halophile *Vibrio natriegens* with a chimera of *Is*PETase and *Is*MHETase from *I. sakaiensis*, with the aim of degrading PET present in seawater. PET was almost completely transformed to TPA, and it was determined that the strain would take 24 years to completely decompose 1 g/L of PET. The creation of chimeras is a novel technique that still has a long way to go in relation to the degradation of synthetic polymers.

Concluding with this polymer, numerous techniques for its depolymerization have been carried out to improve the efficiency of enzymes or modify bacteria with these PETases. Today, research is still needed to obtain more efficient methods for bacteria. This is due to the physical–chemical structure of PET and its own disposition in space, with the semi-crystalline structure being more recalcitrant. The use of enzymes to treat the problem of this polymer has the advantage of being harmless to living beings and the environment, so further research should be carried out.

3.2. Biodegradation of Other Plastic Polymers

Genetic engineering approaches carried out to degrade synthetic polymers have not only focused on PET but also on other polymers, as described below.

3.2.1. Polyethylene (PE)

PE is a particular type of plastic that results from the polymerization of ethylene. A lot of plastics in this group are copolymers of ethylene and olefins, the latter of which result in branches in the plastic's structure. This results in a wide variety of polyethylenes with different properties [89]. Multiple organisms have been discovered that can degrade different types of polyethylene [90–92].

In a study of genetic engineering, Gyung Yoon et al., who expressed the alkane hydroxylase gene (alkB) of *Pseudomonas* sp. E4 in *E. coli* BL21, found that it was able to degrade low-molecular-weight polyethylene (LMWPE) [53]. Kong et al. bound the latex clearing protein derived from *Streptomyces* sp. strain K30 (LcpK30) to the anchor peptide LCI and expressed it in *E. coli* in order to improve its ability to degrade low-density polyethylene (LDPE) [93].

3.2.2. Poly(butylene Adipate-*co*-terephthalate) (PBAT)

PBAT is a long-chain aliphatic–aromatic copolyester formed by subunits of TPA, adipic acid and 1,4-butanediol linked by ester bonds. It is estimated that its global production was more than 360 million tons in 2018 [94]. Its importance relies on the fact that, like PET, it is a plastic used in daily life due to the advantages it presents. It is easy and cheap to produce, resistant and very versatile due to its chemical structure, which gives it the property of being rigid and flexible [95]. PBAT is produced by the condensation of TPA, 1,4-butanediol and adipic acid, and catalysts with tin, zinc and titanium are used [96]. It is widely used in the textile industry, agriculture and packaging. It is also present in organic waste bags because its ester bonds are more easily degradable by enzymes. Its biodegradability has been demonstrated in previous studies, so it is considered a bioplastic [97].

However, its biodegradation rate is low [95]. For this reason, some studies have been carried out with cutinases to better understand the hydrolysis mechanism of PBAT. Due to the presence of ester bonds in PBAT, it is possible to extrapolate the cutinases research carried out on PET to biodegrade PBAT. This happens with Yang et al. [98], where researchers studied the structure of *T. fusca* cutinase TfCut. This enzyme is known to degrade PBAT into its monomers. A double-mutated cutinase, TfCut-DM, was created. TfCut-DM depolymerized all the PBAT into TPA after 48 h of culture, while, at that time, the native enzyme released more 1,4-butanediol-TPA than TPA. On the other hand, genes related to anaerobic living beings, such as *Clostridium botulinum*, have been researched. This is the case with the esterases Cbotu_EstA and Cbotu_EstB, which are expressed heterologously in *E. coli* BL21-Gold(DE3). They have been shown to be able to hydrolyze PBAT. However, some differences in the amino acids of its active site make Cbotu_EstA more efficient than Cbotu_EstB. In silico representations show that the active site of the first protein is larger, which can facilitate the entry of the polymer [49].

3.2.3. Poly(butylene succinate) (PBS)

PBS is an aliphatic polyester formed by the condensation of succinic acid and 1,4-butanediol. In the past, it was produced from compounds derived from petroleum. Nowadays, it is produced from renewable resources, such as sugar cane and corn, through bacterial fermentation [99]. This is a point in its favor in terms of being produced in greater quantities, along with the physical and chemical properties it has. It has a crystalline structure that gives it rigidity and some ductility and a wide temperature range with which to work. Among other uses, it is present in the textile industry and is a possible substitute for polypropylene [100].

The same proteins that hydrolyze PET and PBAT break the ester bonds of PBS. A study on the genetic engineering of cutinases is the one of Gamerith et al. [51], mentioned above for PET. In the article, they used Thc_Cut1 cutinase and two mutants, of which one of them resulted in hydrolyzing the PBS at a higher rate, degrading 92% of the dry weight after 96 h. This confirms its potential for PBS biodegradation.

3.2.4. Polylactic Acid (PLA)

PLA is an aliphatic polyester formed by monomers of lactic acid. In its manufacturing, it is normally made from hydroxyl acids. It is based on bacterial fermentation, as in the case of PBS, to synthesize lactic acid. Later, lactic acid is purified, and the monomers are condensated and linked by ester bonds. PLA is considered a bioplastic that can be chemically depolymerized to lactic acid in recycling processes. PLA is rigid and transparent, and compared to other bioplastics, it has a long durability. Among its uses, it is used as food packaging, for 3D printing, in medicine and the textile industry. PLA can be degraded by cutinases and esterases, as in the case of PET, PBAT and PBS [101].

There are almost no genetic engineering studies focused on PLA, partly because the same enzymes can hydrolyze multiple polymers and partly because PLA is not as recalcitrant as others, such as PET, making PLA less interesting when it comes to developing biodegradation strategies. There is, however, a patent on a protease from *Thermus* sp. Rt41A that has been mutated and whose variants are meant to be used to biologically degrade PLA and recover lactic acid monomers [48].

3.2.5. Poly(3-hydroxybutyrate-*co*-3-hydroxyvalerate) (PHBV)

PHBV, one of the most known PHAs, is a copolymer of hydroxyl-butyrate (HB) and hydroxyl-valerate (HV). When compared to PHB, another well-studied PHA, PHBV has better flexibility and a lower melting temperature [102]. The ratio between both monomers determines the plastic's properties, with a higher content of HB resulting in higher strength and a higher content of HV resulting in higher flexibility and toughness [103]. PHBV's properties can be improved in different ways [104], for example, by mixing it with agro-residues, which results in green composites that have enhanced physical properties [102], or by forming small crystal nuclei and then drawing the material, resulting in fibers with high tensile strength [105]. PHBV has many different uses [106]. It can be used for coatings and packaging [107–109] as well as 3D printing [110]. PHBV can also be used in the textile industry, especially when blended with poly(lactic acid) (PLA), a biobased and biodegradable plastic [111], and in cosmetics as microbeads for exfoliants [112]. The medical field has also found various uses for PHBV. For example, it can be used to make biocompatible scaffolds [113,114] or slowly release drugs like insulin [115] or drugs for treating tumors [116]. Each of the monomers of PHBV is synthesized in a different metabolic pathway. In one pathway, acetyl-CoA is transformed into acetoacetyl-CoA, which, in turn, is transformed into 3-hydroxybutyryl-CoA. In the other pathway, oxaloacetate is sequentially transformed into threonine, 2-ketobutyrate, propionyl-CoA, 3-ketovaleryl-CoA and R-3-Hydroxyvaleryl-CoA [117].

PHBV waste can be recycled. For example, PHBV-based face masks can be chemically recycled through hydrolysis [118]. However, as PHBV is a biodegradable plastic, carrying out its biodegradation is preferable. Its biodegradability when released to the environment is slow and depends greatly on environmental conditions [119], but PHBV has been estimated to biodegrade in rivers [120], and the highest degradation is reached when in activated sludge [121] or compost [122,123]. PHBV-degrading bacteria have been isolated from soil, specifically *Actinomadura* sp. AF-555 [124] and *Streptomyces* sp. MG [125], but very little is known about specific enzymes that can degrade PHBV. In an aforementioned study, the enzymes Thc_Cut1 and Thc_Cut1_koST from *Thermobifida cellulosilytica* DSM44535, expressed in *P. pastoris*, were able to degrade PHBV, though to a very limited extent [51].

3.2.6. Polyvinyl Acetate (PVAC)

PVAC, also known as PVAc, is a petrochemical plastic that is mostly used as an adhesive, particularly for paper and wood [126], but it has also found a use as part of graphene-derived composites [127]. The monomer, vinyl acetate monomer (VAM), is synthesized through the oxidative addition of acetic acid to ethylene, and acetic acid can be, in turn, synthesized from ethylene [128]. Polymerization reactions in PVAC include emulsion-, suspension- and solution polymerization [129].

It has been reported that fungi can grow on items made from PVAC, suggesting its biodegradation [130], but PVAC is generally considered to be non-biodegradable [128]. Liu et al. [50] developed a chimeric lipase–cutinase (Lip-Cut), originally *from Thermomyces lanuginosus* and *Thielavia terrestris* NRRL 8126, respectively, and expressed it in *P. pastoris.* They studied the use of this Lip-Cut in the process of deinking waste paper and found that the synergy between both enzymes improves the ability of the cutinase to both degrade PVAC and deink paper. Also, regarding paper waste treatment, Liu et al. [131] fused the cutinase from *Humicola insolens* to the *E. coli* anchor peptide OMP25, which can be adsorbed on the polyester polymethylmethacrylate (PMMA). They expressed the fusion protein in *P. pastoris* and concluded that the anchor peptide enhances the cutinase's ability to degrade PVAC and the so-called stickies in waste paper recycling.

3.2.7. Poly(ε-caprolactone) (PCL)

PCL is a plastic well known for its use in the biomedical field [132,133], as it is biocompatible [134]. It can be used in drug delivery or wound dressing [135], but it stands out as a scaffold for tissue engineering, whether on its own [136] or as part of a blend or composite material [137]. Its monomer is ε-caprolactone, and its polymerization can be carried out by opening the monomer's ring followed by the addition or, alternatively, by polymerizing 6-hydroxycaproic acid [138].

PCL is a biodegradable plastic [139,140], and its degradation is especially high when in anaerobic sludge [141] and compost [142] or when performed by bacteria isolated from compost [143]. Regarding engineered bacteria to degrade PCL, two examples have already been mentioned. Sulaiman et al. [54] found that LC-cutinase could degrade PCL in addition to, and better than, PET. The aforementioned CUT190 cutinase from *Saccharomonospora viridis* AHK190, when double-mutated and expressed in *E. coli* Rosetta-gami B (DE3), had an improved degradation activity of PCL, among other polymers [58].

3.2.8. Polyurethane (PU or PUR)

PU is a polymer synthesized by the union of a diol and a diisocyanate molecule linked by amide bonds. To accelerate the process, catalysts, such as 1,4-diazabicyclo[2.2.2]octane or dibutyltin dilaurate, can be added. Polyurethane is the eighth most consumed polymer in the world, with 18.6 MMt/yr. Its structure can be highly varied, and it has a lot of uses. Its main use is as a foaming agent, but it is also manufactured in thermal insulation processes and to protect against abrasion and corrosion, among other uses. Its drawback is that this manufacturing requires toxic substrates, making it difficult to recycle. For this reason, alternative routes to synthesize it are being sought. In addition, using biodegradation to eliminate this polymer in a harmless way is being studied [144]. PU has amide bonds that are hydrolyzed by amidases, releasing the monomers. To increase the degradation of PU, some researchers improved the adsorption of an amidase from *Nocardia farcinica* to PU by fusing it to a polymer-binding module originally from a polyhydroxyalkanoate depolymerase obtained from *Alcaligenes faecalis* [52]. However, there are not many studies based on the biodegradability of PU.

4. Conclusions

The magnitude of plastic usage and the later production of its waste, which in most cases remains in nature for decades, makes the search for alternative bioplastic materials or other compounds a matter of major importance for today's society. In addition, the biodegradation of these materials and, where possible, the use of their constitutive monomers as carbon and energy sources for the growth of microorganisms that can produce valuable compounds are important goals to achieve [145].

Despite the fact that there are a large number of studies based on genetic modifications, most of them were carried out through modifications in the amino acid sequence. New strategies have been recently tested, such as the use of chimeras or fusion proteins, among others, but these approaches are a minority. Further research in the use of genetic techniques, as well as combining different strategies, is necessary to obtain new strains or proteins with a greater degradative capacity of these synthetic polymers.

An example of a new research field is the use of transcriptional regulators, such as Garrido et al. [146] have carried out. This work tested if the FarA transcription factor in *Aspergillus oryzae* was directly related to the synthesis of CutL1 and HsbA, a cutinase and a hydrophobic surface binding protein that participate in the degradation of poly(butylene succinate-*co*-adipate) (PBSA), a biodegradable polymer. The deletion of FarA produced a mutant with minimal concentrations of *cutL1* and *hsbA* compared with its native strain and without the ability to degrade PBSA. This study differs from the ones described in the main text, as it consists of the basic research of an organism and its genetic regulation. This does not mean that the other studies ignore the subject of gene expression. In contrast, modifying bacteria to express an enzyme requires careful consideration of the genetic elements necessary for its successful production. The difference relies on studying genetic expression in the native organism, which cannot only help us understand their genetic regulation and metabolism but also give us hints on different approaches to enhance plastic degradation and how to use the native organism instead of relying on a different host organism.

It is also relevant to point out that the research tends to focus on enhancing an enzyme's ability to degrade specific polymers, with almost no studies evaluating how the modifications performed may change specificity toward other polymers. Limiting the analysis to just a few substrates—for which the enzyme is already specific—is comprehensible, but using a wider variety of polymers could indicate if the modifications have a greater effect than expected and may be more useful.

In addition, there is little work on the tridimensional structure of the different enzymes—native or modified—that have been mentioned. Deeper research is necessary on this matter in order to better understand why changes happen regarding the affinity for the substrate or the stability, among other characteristics, according to the exchanged residues. Obtaining the structure of enzymes can be achieved not only through X-ray crystallography but also in silico by modeling. This is especially useful when proteins cannot be successfully purified and crystallized, but modeling goes even further. From molecular dynamics to substrate docking, these tools are used in some of the referenced studies to analyze enzymes' properties and enhance their activities.

Similar to plastics themselves, approaches to studying their biodegradation are diverse, and this variety often comes with a lack of consistency between studies. The conditions in which the enzymes are used can differ, and some studies do not calculate enzymatic activity and limit themselves to detecting solid plastic disintegration or monomer liberation, and the works that calculate enzymatic activity can use different units. This can sometimes make it complicated to compare their conclusions.

Overall, more research is required to develop effective, i.e., quicker, safer and more efficient, biodegradation strategies for plastics, if possible. This applies not just to the plastics that are briefly addressed in this work but also to all plastics in general. Microbial biotechnology and genetic engineering approaches, together with the current development of artificial intelligence tools that provide a new direction in the study and design of novel of enzymes, can facilitate the generation and optimization of several types of plastic-degrading enzymes and valorization processes.

Nowadays, we are getting closer to achieving this aim thanks to the latest advances in DNA sequencing, metagenomics, bioinformatics, genome mining and machine learning tools, in conjunction with new genetic engineering techniques, such as CRISPR-Cas technologies.

Author Contributions: All authors participated in the writing of the work. F.S.-B. edited and handled the manuscript. All authors have read and agreed to the published version of the manuscript.

Funding: This research received no external funding.

Acknowledgments: We acknowledge Laura Andrea Silva López for helping with the representation of the molecular structures shown in Figures 3A and 4. Fundación Ramón Areces (CLU 2017-09 and ProgStrain) is also gratefully acknowledged for supporting the contract of Carlos de la Fuente Tagarro.

Conflicts of Interest: The authors declare no conflicts of interest.

References

1. Shrivastava, A. Introduction to Plastics Engineering. In *Introduction to Plastics Engineering*; Elsevier: Amsterdam, The Netherlands, 2018; pp. 1–16. ISBN 9780323395007.
2. Nayanathara Thathsarani Pilapitiya, P.G.C.; Ratnayake, A.S. The World of Plastic Waste: A Review. *Clean. Mater.* **2024**, *11*, 100220. [CrossRef]
3. Geyer, R.; Jambeck, J.R.; Law, K.L. Production, Use, and Fate of All Plastics Ever Made. *Sci. Adv.* **2017**, *3*, e1700782. [CrossRef] [PubMed]
4. Gambarini, V.; Pantos, O.; Kingsbury, J.M.; Weaver, L.; Handley, K.M.; Lear, G. PlasticDB: A Database of Microorganisms and Proteins Linked to Plastic Biodegradation. *Database* **2022**, *2022*, baac008. [CrossRef]
5. Tournier, V.; Duquesne, S.; Guillamot, F.; Cramail, H.; Taton, D.; Marty, A.; André, I. Enzymes' Power for Plastics Degradation. *Chem. Rev.* **2023**, *123*, 5612–5701. [CrossRef]
6. Zhou, S.; Kuester, T.; Bochow, M.; Bohn, N.; Brell, M.; Kaufmann, H. A Knowledge-Based, Validated Classifier for the Identification of Aliphatic and Aromatic Plastics by WorldView-3 Satellite Data. *Remote Sens. Environ.* **2021**, *264*, 112598. [CrossRef]
7. Niaounakis, M. (Ed.) *Management of Marine Plastic Debris*; Plastics Design Library; William Andrew Publishing: Oxford, UK, 2017; ISBN 9780323443548.

8. UN Environment Programme. Drowning in Plastics—Marine Litter and Plastic Waste Vital Graphics. *UNEP—UN Environment Programme*. Available online: http://www.unep.org/resources/report/drowning-plastics-marine-litter-and-plastic-waste-vital-graphics (accessed on 23 March 2024).
9. Standard Practice for Coding Plastic Manufactured Articles for Resin Identification. Available online: https://www.astm.org/d7611_d7611m-21.html (accessed on 23 March 2024).
10. Plastics Europe. Plastics—The Fast Facts 2023. Available online: https://plasticseurope.org/knowledge-hub/plastics-the-fast-facts-2023/ (accessed on 23 March 2024).
11. Plastics Europe. Plastics—The Facts 2021. Available online: https://plasticseurope.org/knowledge-hub/plastics-the-facts-2021/ (accessed on 23 March 2024).
12. Institute, E. Home. Statistical Review of World Energy. Available online: https://www.energyinst.org/statistical-review/home (accessed on 23 March 2024).
13. *Plastic Atlas 2019: Facts and Figures about the World of Synthetic Polymers*; Heinrich Böll Foundation: Berlin, Germany, 2019; ISBN 9783869282114.
14. Jambeck, J.R.; Geyer, R.; Wilcox, C.; Siegler, T.R.; Perryman, M.; Andrady, A.; Narayan, R.; Law, K.L. Plastic Waste Inputs from Land into the Ocean. *Science* **2015**, *347*, 768–771. [CrossRef]
15. Chamas, A.; Moon, H.; Zheng, J.; Qiu, Y.; Tabassum, T.; Jang, J.H.; Abu-Omar, M.; Scott, S.L.; Suh, S. Degradation Rates of Plastics in the Environment. *ACS Sustain. Chem. Eng.* **2020**, *8*, 3494–3511. [CrossRef]
16. United States Department of Commerce, National Oceanic and Atmospheric Administration. Ingestion 2014—Report on the Occurrence and Health Effects of Anthropogenic Debris Ingested by Marine Organisms. Available online: https://marinedebris.noaa.gov/wildlife-and-habitat-impacts/occurrence-and-health-effects-anthropogenic-debris-ingested-marine-organisms (accessed on 23 March 2024).
17. Environment, U.N. From Pollution to Solution: A Global Assessment of Marine Litter and Plastic Pollution. Available online: http://www.unep.org/resources/pollution-solution-global-assessment-marine-litter-and-plastic-pollution (accessed on 23 March 2024).
18. De Souza Machado, A.A.; Kloas, W.; Zarfl, C.; Hempel, S.; Rillig, M.C. Microplastics as an Emerging Threat to Terrestrial Ecosystems. *Glob. Chang. Biol.* **2018**, *24*, 1405–1416. [CrossRef] [PubMed]
19. Wright, S.L.; Kelly, F.J. Plastic and Human Health: A Micro Issue? *Environ. Sci. Technol.* **2017**, *51*, 6634–6647. [CrossRef]
20. Prata, J.C.; Da Costa, J.P.; Lopes, I.; Duarte, A.C.; Rocha-Santos, T. Environmental Exposure to Microplastics: An Overview on Possible Human Health Effects. *Sci. Total Environ.* **2020**, *702*, 134455. [CrossRef]
21. Hahladakis, J.N.; Velis, C.A.; Weber, R.; Iacovidou, E.; Purnell, P. An Overview of Chemical Additives Present in Plastics: Migration, Release, Fate and Environmental Impact during Their Use, Disposal and Recycling. *J. Hazard. Mater.* **2018**, *344*, 179–199. [CrossRef]
22. Hermabessiere, L.; Dehaut, A.; Paul-Pont, I.; Lacroix, C.; Jezequel, R.; Soudant, P.; Duflos, G. Occurrence and Effects of Plastic Additives on Marine Environments and Organisms: A Review. *Chemosphere* **2017**, *182*, 781–793. [CrossRef]
23. Chamanee, G.; Sewwandi, M.; Wijesekara, H.; Vithanage, M. Global Perspective on Microplastics in Landfill Leachate; Occurrence, Abundance, Characteristics, and Environmental Impact. *Waste Manag.* **2023**, *171*, 10–25. [CrossRef] [PubMed]
24. GodvinSharmila, V.; Shanmugavel, S.P.; Tyagi, V.K.; Rajesh Banu, J. Microplastics as Emergent Contaminants in Landfill Leachate: Source, Potential Impact and Remediation Technologies. *J. Environ. Manag.* **2023**, *343*, 118240. [CrossRef]
25. Redko, V.; Wolska, L.; Potrykus, M.; Olkowska, E.; Cieszyńska-Semenowicz, M.; Tankiewicz, M. Environmental Impacts of 5-Year Plastic Waste Deposition on Municipal Waste Landfills: A Follow-up Study. *Sci. Total Environ.* **2024**, *906*, 167710. [CrossRef]
26. Ali, S.S.; Elsamahy, T.; Koutra, E.; Kornaros, M.; El-Sheekh, M.; Abdelkarim, E.A.; Zhu, D.; Sun, J. Degradation of Conventional Plastic Wastes in the Environment: A Review on Current Status of Knowledge and Future Perspectives of Disposal. *Sci. Total Environ.* **2021**, *771*, 144719. [CrossRef] [PubMed]
27. European Bioplastics. What Are Bioplastics? Available online: https://www.european-bioplastics.org/bioplastics/ (accessed on 23 March 2024).
28. Raza, Z.A.; Abid, S.; Banat, I.M. Polyhydroxyalkanoates: Characteristics, Production, Recent Developments and Applications. *Int. Biodeterior. Biodegrad.* **2018**, *126*, 45–56. [CrossRef]
29. Koller, M.; Salerno, A.; Braunegg, G. Polyhydroxyalkanoates: Basics, Production and Applications of Microbial Biopolyesters. In *Bio-Based Plastics*; Kabasci, S., Ed.; Wiley: Hoboken, NJ, USA, 2013; pp. 137–170. ISBN 9781119994008.
30. Li, Z.; Yang, J.; Loh, X.J. Polyhydroxyalkanoates: Opening Doors for a Sustainable Future. *NPG Asia Mater.* **2016**, *8*, e265. [CrossRef]
31. Bishop, G.; Styles, D.; Lens, P.N.L. Environmental Performance of Bioplastic Packaging on Fresh Food Produce: A Consequential Life Cycle Assessment. *J. Clean. Prod.* **2021**, *317*, 128377. [CrossRef]
32. Manfra, L.; Marengo, V.; Libralato, G.; Costantini, M.; De Falco, F.; Cocca, M. Biodegradable Polymers: A Real Opportunity to Solve Marine Plastic Pollution? *J. Hazard. Mater.* **2021**, *416*, 125763. [CrossRef]
33. Atiwesh, G.; Mikhael, A.; Parrish, C.C.; Banoub, J.; Le, T.-A.T. Environmental Impact of Bioplastic Use: A Review. *Heliyon* **2021**, *7*, e07918. [CrossRef]
34. Emadian, S.M.; Onay, T.T.; Demirel, B. Biodegradation of Bioplastics in Natural Environments. *Waste Manag.* **2017**, *59*, 526–536. [CrossRef] [PubMed]

35. García-Depraect, O.; Bordel, S.; Lebrero, R.; Santos-Beneit, F.; Börner, R.A.; Börner, T.; Muñoz, R. Inspired by Nature: Microbial Production, Degradation and Valorization of Biodegradable Bioplastics for Life-Cycle-Engineered Products. *Biotechnol. Adv.* **2021**, *53*, 107772. [CrossRef] [PubMed]
36. Nandakumar, A.; Chuah, J.-A.; Sudesh, K. Bioplastics: A Boon or Bane? *Renew. Sustain. Energy Rev.* **2021**, *147*, 111237. [CrossRef]
37. Savva, K.; Borrell, X.; Moreno, T.; Pérez-Pomeda, I.; Barata, C.; Llorca, M.; Farré, M. Cytotoxicity Assessment and Suspected Screening of Plastic Additives in Bioplastics of Single-Use Household Items. *Chemosphere* **2023**, *313*, 137494. [CrossRef]
38. Di Cesare, A.; Pinnell, L.J.; Brambilla, D.; Elli, G.; Sabatino, R.; Sathicq, M.B.; Corno, G.; O'Donnell, C.; Turner, J.W. Bioplastic Accumulates Antibiotic and Metal Resistance Genes in Coastal Marine Sediments. *Environ. Pollut.* **2021**, *291*, 118161. [CrossRef] [PubMed]
39. Cucina, M.; De Nisi, P.; Tambone, F.; Adani, F. The Role of Waste Management in Reducing Bioplastics' Leakage into the Environment: A Review. *Bioresour. Technol.* **2021**, *337*, 125459. [CrossRef]
40. *UNE-EN 17033:2018*; Plastics—Biodegradable Mulch Films for Use in Agriculture and Horticulture—Requirements and Test Methods. UNE: Madrid, Spain, 2018. Available online: https://www.une.org/encuentra-tu-norma/busca-tu-norma/norma?c=N0060868 (accessed on 23 March 2024).
41. *UNE-EN 14995:2007*; Plastics—Evaluation of Compostability—Test Scheme and Specifications. UNE: Madrid, Spain, 2018. Available online: https://www.une.org/encuentra-tu-norma/busca-tu-norma/norma?c=N0039863 (accessed on 23 March 2024).
42. *UNE-EN 13432:2001*; Requirements for Packaging Recoverable through Composting and Biodegradation. Test Scheme and Evaluation Criteria for the Final Acceptance of Packaging. UNE: Madrid, Spain, 2018. Available online: https://www.une.org/encuentra-tu-norma/busca-tu-norma/norma?c=N0024465 (accessed on 23 March 2024).
43. *ISO 17088:2021*; Plastics—Organic Recycling—Specifications for Compostable Plastics. ISO: Geneva, Switzerland, 2021. Available online: https://www.iso.org/standard/74994.html (accessed on 23 March 2024).
44. *ISO 22403:2020*; Plastics—Assessment of the Intrinsic Biodegradability of Materials Exposed to Marine Inocula under Mesophilic Aerobic Laboratory Conditions—Test Methods and Requirements. ISO: Geneva, Switzerland, 2020. Available online: https://www.iso.org/es/contents/data/standard/07/31/73121.html (accessed on 23 March 2024).
45. *ISO 17556:2019*; Plastics—Determination of the Ultimate Aerobic Biodegradability of Plastic Materials in Soil by Measuring the Oxygen Demand in a Respirometer or the Amount of Carbon Dioxide Evolved. ISO: Geneva, Switzerland, 2019. Available online: https://www.iso.org/es/contents/data/standard/07/49/74993.html (accessed on 23 March 2024).
46. *ISO 14853:2016*; Plastics—Determination of the Ultimate Anaerobic Biodegradation of Plastic Materials in an Aqueous System—Method by Measurement of Biogas Production. ISO: Geneva, Switzerland, 2016. Available online: https://www.iso.org/es/contents/data/standard/06/78/67804.html (accessed on 23 March 2024).
47. Chia, W.Y.; Ying Tang, D.Y.; Khoo, K.S.; Kay Lup, A.N.; Chew, K.W. Nature's Fight against Plastic Pollution: Algae for Plastic Biodegradation and Bioplastics Production. *Environ. Sci. Ecotechnology* **2020**, *4*, 100065. [CrossRef] [PubMed]
48. Marty, A.; Duquense, S.; Guicherd, M.; Vuillemin, M.; Ben Khaled, M. Improved Plastic Degrading Proteases. WO 2018/1091183 A1, 21 June 2018.
49. Perz, V.; Baumschlager, A.; Bleymaier, K.; Zitzenbacher, S.; Hromic, A.; Steinkellner, G.; Pairitsch, A.; Łyskowski, A.; Gruber, K.; Sinkel, C.; et al. Hydrolysis of Synthetic Polyesters by *Clostridium botulinum* Esterases. *Biotech Bioeng.* **2016**, *113*, 1024–1034. [CrossRef]
50. Liu, M.; Yang, S.; Long, L.; Cao, Y.; Ding, S. Engineering a Chimeric Lipase-Cutinase (Lip-Cut) for Efficient Enzymatic Deinking of Waste Paper. *BioResources* **2017**, *13*, 981–996. [CrossRef]
51. Gamerith, C.; Vastano, M.; Ghorbanpour, S.M.; Zitzenbacher, S.; Ribitsch, D.; Zumstein, M.T.; Sander, M.; Herrero Acero, E.; Pellis, A.; Guebitz, G.M. Enzymatic Degradation of Aromatic and Aliphatic Polyesters by *P. pastoris* Expressed Cutinase 1 from *Thermobifida cellulosilytica*. *Front. Microbiol.* **2017**, *8*, 938. [CrossRef] [PubMed]
52. Gamerith, C.; Herrero Acero, E.; Pellis, A.; Ortner, A.; Vielnascher, R.; Luschnig, D.; Zartl, B.; Haernvall, K.; Zitzenbacher, S.; Strohmeier, G.; et al. Improving Enzymatic Polyurethane Hydrolysis by Tuning Enzyme Sorption. *Polym. Degrad. Stab.* **2016**, *132*, 69–77. [CrossRef]
53. Gyung Yoon, M.; Jeong Jeon, H.; Nam Kim, M. Biodegradation of Polyethylene by a Soil Bacterium and AlkB Cloned Recombinant Cell. *J. Bioremed. Biodegrad.* **2012**, *3*, 1–8. [CrossRef]
54. Sulaiman, S.; Yamato, S.; Kanaya, E.; Kim, J.-J.; Koga, Y.; Takano, K.; Kanaya, S. Isolation of a Novel Cutinase Homolog with Polyethylene Terephthalate-Degrading Activity from Leaf-Branch Compost by Using a Metagenomic Approach. *Appl. Environ. Microbiol.* **2012**, *78*, 1556–1562. [CrossRef]
55. Araújo, R.; Silva, C.; O'Neill, A.; Micaelo, N.; Guebitz, G.; Soares, C.M.; Casal, M.; Cavaco-Paulo, A. Tailoring Cutinase Activity towards Polyethylene Terephthalate and Polyamide 6,6 Fibers. *J. Biotechnol.* **2007**, *128*, 849–857. [CrossRef] [PubMed]
56. Silva, C.; Da, S.; Silva, N.; Matamá, T.; Araújo, R.; Martins, M.; Chen, S.; Chen, J.; Wu, J.; Casal, M.; et al. Engineered *Thermobifida fusca* Cutinase with Increased Activity on Polyester Substrates. *Biotechnol. J.* **2011**, *6*, 1230–1239. [CrossRef] [PubMed]
7. Herrero Acero, E.; Ribitsch, D.; Dellacher, A.; Zitzenbacher, S.; Marold, A.; Steinkellner, G.; Gruber, K.; Schwab, H.; Guebitz, G.M. Surface Engineering of a Cutinase from *Thermobifida cellulosilytica* for Improved Polyester Hydrolysis. *Biotechnol. Bioeng.* **2013**, *110*, 2581–2590. [CrossRef] [PubMed]

58. Kawai, F.; Oda, M.; Tamashiro, T.; Waku, T.; Tanaka, N.; Yamamoto, M.; Mizushima, H.; Miyakawa, T.; Tanokura, M. A Novel Ca^{2+}-Activated, Thermostabilized Polyesterase Capable of Hydrolyzing Polyethylene Terephthalate from *Saccharomonospora viridis* AHK190. *Appl. Microbiol. Biotechnol.* **2014**, *98*, 10053–10064. [CrossRef] [PubMed]
59. Wei, R.; Oeser, T.; Schmidt, J.; Meier, R.; Barth, M.; Then, J.; Zimmermann, W. Engineered Bacterial Polyester Hydrolases Efficiently Degrade Polyethylene Terephthalate Due to Relieved Product Inhibition. *Biotech. Bioeng.* **2016**, *113*, 1658–1665. [CrossRef]
60. Barth, M.; Honak, A.; Oeser, T.; Wei, R.; Belisário-Ferrari, M.R.; Then, J.; Schmidt, J.; Zimmermann, W. A Dual Enzyme System Composed of a Polyester Hydrolase and a Carboxylesterase Enhances the Biocatalytic Degradation of Polyethylene Terephthalate Films. *Biotechnol. J.* **2016**, *11*, 1082–1087. [CrossRef]
61. Han, X.; Liu, W.; Huang, J.-W.; Ma, J.; Zheng, Y.; Ko, T.-P.; Xu, L.; Cheng, Y.-S.; Chen, C.-C.; Guo, R.-T. Structural Insight into Catalytic Mechanism of PET Hydrolase. *Nat. Commun.* **2017**, *8*, 2106. [CrossRef] [PubMed]
62. Shirke, A.N.; White, C.; Englaender, J.A.; Zwarycz, A.; Butterfoss, G.L.; Linhardt, R.J.; Gross, R.A. Stabilizing Leaf and Branch Compost Cutinase (LCC) with Glycosylation: Mechanism and Effect on PET Hydrolysis. *Biochemistry* **2018**, *57*, 1190–1200. [CrossRef] [PubMed]
63. Joo, S.; Cho, I.J.; Seo, H.; Son, H.F.; Sagong, H.-Y.; Shin, T.J.; Choi, S.Y.; Lee, S.Y.; Kim, K.-J. Structural Insight into Molecular Mechanism of Poly(Ethylene Terephthalate) Degradation. *Nat. Commun.* **2018**, *9*, 382. [CrossRef] [PubMed]
64. Liu, B.; He, L.; Wang, L.; Li, T.; Li, C.; Liu, H.; Luo, Y.; Bao, R. Protein Crystallography and Site-Direct Mutagenesis Analysis of the Poly(Ethylene Terephthalate) Hydrolase PETase from *Ideonella sakaiensis*. *ChemBioChem* **2018**, *19*, 1471–1475. [CrossRef] [PubMed]
65. Austin, H.P.; Allen, M.D.; Donohoe, B.S.; Rorrer, N.A.; Kearns, F.L.; Silveira, R.L.; Pollard, B.C.; Dominick, G.; Duman, R.; El Omari, K.; et al. Characterization and Engineering of a Plastic-Degrading Aromatic Polyesterase. *Proc. Natl. Acad. Sci. USA* **2018**, *115*, E4350–E4357. [CrossRef]
66. Ma, Y.; Yao, M.; Li, B.; Ding, M.; He, B.; Chen, S.; Zhou, X.; Yuan, Y. Enhanced Poly(Ethylene Terephthalate) Hydrolase Activity by Protein Engineering. *Engineering* **2018**, *4*, 888–893. [CrossRef]
67. Son, H.F.; Cho, I.J.; Joo, S.; Seo, H.; Sagong, H.-Y.; Choi, S.Y.; Lee, S.Y.; Kim, K.-J. Rational Protein Engineering of Thermo-Stable PETase from *Ideonella sakaiensis* for Highly Efficient PET Degradation. *ACS Catal.* **2019**, *9*, 3519–3526. [CrossRef]
68. Yan, F.; Wei, R.; Cui, Q.; Bornscheuer, U.T.; Liu, Y. Thermophilic Whole-cell Degradation of Polyethylene Terephthalate Using Engineered *Clostridium thermocellum*. *Microb. Biotechnol.* **2021**, *14*, 374–385. [CrossRef]
69. Tournier, V.; Topham, C.M.; Gilles, A.; David, B.; Folgoas, C.; Moya-Leclair, E.; Kamionka, E.; Desrousseaux, M.-L.; Texier, H.; Gavalda, S.; et al. An Engineered PET Depolymerase to Break down and Recycle Plastic Bottles. *Nature* **2020**, *580*, 216–219. [CrossRef]
70. Sadler, J.C.; Wallace, S. Microbial Synthesis of Vanillin from Waste Poly(Ethylene Terephthalate). *Green Chem.* **2021**, *23*, 4665–4672. [CrossRef]
71. Zhu, B.; Ye, Q.; Seo, Y.; Wei, N. Enzymatic Degradation of Polyethylene Terephthalate Plastics by Bacterial Curli Display PETase. *Environ. Sci. Technol. Lett.* **2022**, *9*, 650–657. [CrossRef]
72. Chen, Y.; Zhang, S.; Zhai, Z.; Zhang, S.; Ma, J.; Liang, X.; Li, Q. Construction of Fusion Protein with Carbohydrate-Binding Module and Leaf-Branch Compost Cutinase to Enhance the Degradation Efficiency of Polyethylene Terephthalate. *IJMS* **2023**, *24*, 2780. [CrossRef] [PubMed]
73. Li, T.; Menegatti, S.; Crook, N. Breakdown of polyethylene therepthalate Microplastics under Saltwater Conditions Using Engineered *Vibrio natriegens*. *AIChE J.* **2023**, *69*, e18228. [CrossRef]
74. Ellis, L.D.; Rorrer, N.A.; Sullivan, K.P.; Otto, M.; McGeehan, J.E.; Román-Leshkov, Y.; Wierckx, N.; Beckham, G.T. Chemical and Biological Catalysis for Plastics Recycling and Upcycling. *Nat. Catal.* **2021**, *4*, 539–556. [CrossRef]
75. Pang, K.; Kotek, R.; Tonelli, A. Review of Conventional and Novel Polymerization Processes for Polyesters. *Prog. Polym. Sci.* **2006**, *31*, 1009–1037. [CrossRef]
76. Papadopoulou, A.; Hecht, K.; Buller, R. Enzymatic PET Degradation. *Chimia* **2019**, *73*, 743. [CrossRef]
77. Dhandapani, A.; Krishnasamy, S.; Thiagamani, S.M.K.; Periasamy, D.; Muthukumar, C.; Sundaresan, T.K.; Ali, S.; Kurniawan, R. Evolution, Prospects, and Predicaments of Polymers in Marine Applications: A Potential Successor to Traditional Materials. *Recycling* **2024**, *9*, 8. [CrossRef]
78. Muringayil Joseph, T.; Azat, S.; Ahmadi, Z.; Moini Jazani, O.; Esmaeili, A.; Kianfar, E.; Haponiuk, J.; Thomas, S. Polyethylene Terephthalate (PET) Recycling: A Review. *Case Stud. Chem. Environ. Eng.* **2024**, *9*, 100673. [CrossRef]
79. Kushwaha, A.; Goswami, L.; Singhvi, M.; Kim, B.S. Biodegradation of Poly(Ethylene Terephthalate): Mechanistic Insights, Advances, and Future Innovative Strategies. *Chem. Eng. J.* **2023**, *457*, 141230. [CrossRef]
80. Lv, S.; Li, Y.; Zhao, S.; Shao, Z. Biodegradation of Typical Plastics: From Microbial Diversity to Metabolic Mechanisms. *IJMS* **2024**, *25*, 593. [CrossRef]
81. Yoshida, S.; Hiraga, K.; Takehana, T.; Taniguchi, I.; Yamaji, H.; Maeda, Y.; Toyohara, K.; Miyamoto, K.; Kimura, Y.; Oda, K. A Bacterium That Degrades and Assimilates Poly(Ethylene Terephthalate). *Science* **2016**, *351*, 1196–1199. [CrossRef] [PubMed]
82. Magalhães, R.P.; Cunha, J.M.; Sousa, S.F. Perspectives on the Role of Enzymatic Biocatalysis for the Degradation of Plastic PET. *IJMS* **2021**, *22*, 11257. [CrossRef] [PubMed]
83. Li, A.; Sheng, Y.; Cui, H.; Wang, M.; Wu, L.; Song, Y.; Yang, R.; Li, X.; Huang, H. Discovery and Mechanism-Guided Engineering of BHET Hydrolases for Improved PET Recycling and Upcycling. *Nat. Commun.* **2023**, *14*, 4169. [CrossRef] [PubMed]

84. Kalathil, S.; Miller, M.; Reisner, E. Microbial Fermentation of Polyethylene Terephthalate (PET) Plastic Waste for the Production of Chemicals or Electricity. *Angew. Chem. Int. Ed.* **2022**, *61*, e202211057. [CrossRef] [PubMed]
85. Diao, J.; Hu, Y.; Tian, Y.; Carr, R.; Moon, T.S. Upcycling of Poly(Ethylene Terephthalate) to Produce High-Value Bio-Products. *Cell Rep.* **2023**, *42*, 111908. [CrossRef] [PubMed]
86. Matak, M.Y.; Moghaddam, M.E. The Role of Short-Range Cys171–Cys178 Disulfide Bond in Maintaining Cutinase Active Site Integrity: A Molecular Dynamics Simulation. *Biochem. Biophys. Res. Commun.* **2009**, *390*, 201–204. [CrossRef] [PubMed]
87. Han, W.; Zhang, J.; Chen, Q.; Xie, Y.; Zhang, M.; Qu, J.; Tan, Y.; Diao, Y.; Wang, Y.; Zhang, Y. Biodegradation of Poly(Ethylene Terephthalate) through PETase Surface-Display: From Function to Structure. *J. Hazard. Mater.* **2024**, *461*, 132632. [CrossRef] [PubMed]
88. Müller, R.; Schrader, H.; Profe, J.; Dresler, K.; Deckwer, W. Enzymatic Degradation of Poly(Ethylene Terephthalate): Rapid Hydrolyse Using a Hydrolase from *T. fusca*. *Macromol. Rapid Commun.* **2005**, *26*, 1400–1405. [CrossRef]
89. Kissin, Y.V. *Polyethylene: End-Use Properties and Their Physical Meaning*; Carl Hanser Verlag GmbH & Co. KG: München, Germany, 2020; ISBN 9781569908310.
90. Zhang, Y.; Pedersen, J.N.; Eser, B.E.; Guo, Z. Biodegradation of Polyethylene and Polystyrene: From Microbial Deterioration to Enzyme Discovery. *Biotechnol. Adv.* **2022**, *60*, 107991. [CrossRef]
91. Nedi, E.G.; Ebu, S.M.; Somboo, M. Identification and Molecular Characterization of Polyethylene Degrading Bacteria from Garbage Dump Sites in Adama, Ethiopia. *Environ. Technol. Innov.* **2024**, *33*, 103441. [CrossRef]
92. Li, X.; Li, G.; Wang, J.; Li, X.; Yang, Y.; Song, D. Elucidating Polyethylene Microplastic Degradation Mechanisms and Metabolic Pathways via Iron-Enhanced Microbiota Dynamics in Marine Sediments. *J. Hazard. Mater.* **2024**, *466*, 133655. [CrossRef] [PubMed]
93. Kong, D.; Zhang, H.; Yuan, Y.; Wu, J.; Liu, Z.; Chen, S.; Zhang, F.; Wang, L. Enhanced Biodegradation Activity toward Polyethylene by Fusion Protein of Anchor Peptide and *Streptomyces* Sp. Strain K30 Latex Clearing Protein. *Int. J. Biol. Macromol.* **2024**, *264*, 130378. [CrossRef] [PubMed]
94. Jehanno, C.; Alty, J.W.; Roosen, M.; De Meester, S.; Dove, A.P.; Chen, E.Y.-X.; Leibfarth, F.A.; Sardon, H. Critical Advances and Future Opportunities in Upcycling Commodity Polymers. *Nature* **2022**, *603*, 803–814. [CrossRef] [PubMed]
95. Shah, A.A.; Kato, S.; Shintani, N.; Kamini, N.R.; Nakajima-Kambe, T. Microbial Degradation of Aliphatic and Aliphatic-Aromatic Co-Polyesters. *Appl. Microbiol. Biotechnol.* **2014**, *98*, 3437–3447. [CrossRef] [PubMed]
96. Jian, J.; Xiangbin, Z.; Xianbo, H. An Overview on Synthesis, Properties and Applications of Poly(Butylene-Adipate-Co-Terephthalate)–PBAT. *Adv. Ind. Eng. Polym. Res.* **2020**, *3*, 19–26. [CrossRef]
97. Witt, U.; Yamamoto, M.; Seeliger, U.; Müller, R.-J.; Warzelhan, V. Biodegradable Polymeric Materials—Not the Origin but the Chemical Structure Determines Biodegradability. *Angew. Chem. Int. Ed.* **1999**, *38*, 1438–1442. [CrossRef]
98. Yang, Y.; Min, J.; Xue, T.; Jiang, P.; Liu, X.; Peng, R.; Huang, J.-W.; Qu, Y.; Li, X.; Ma, N.; et al. Complete Bio-Degradation of Poly(Butylene Adipate-Co-Terephthalate) via Engineered Cutinases. *Nat. Commun.* **2023**, *14*, 1645. [CrossRef] [PubMed]
99. Xu, J.; Guo, B. Poly(Butylene Succinate) and Its Copolymers: Research, Development and Industrialization. *Biotechnol. J.* **2010**, *5*, 1149–1163. [CrossRef]
100. Aliotta, L.; Seggiani, M.; Lazzeri, A.; Gigante, V.; Cinelli, P. A Brief Review of Poly (Butylene Succinate) (PBS) and Its Main Copolymers: Synthesis, Blends, Composites, Biodegradability, and Applications. *Polymers* **2022**, *14*, 844. [CrossRef]
101. Taib, N.-A.A.B.; Rahman, M.R.; Huda, D.; Kuok, K.K.; Hamdan, S.; Bakri, M.K.B.; Julaihi, M.R.M.B.; Khan, A. A Review on Poly Lactic Acid (PLA) as a Biodegradable Polymer. *Polym. Bull.* **2023**, *80*, 1179–1213. [CrossRef]
102. Ahankari, S.S.; Mohanty, A.K.; Misra, M. Mechanical Behaviour of Agro-Residue Reinforced Poly(3-Hydroxybutyrate-Co-3-Hydroxyvalerate), (PHBV) Green Composites: A Comparison with Traditional Polypropylene Composites. *Compos. Sci. Technol.* **2011**, *71*, 653–657. [CrossRef]
103. Inamuddin; Altalhi, T. (Eds.) *Handbook of Bioplastics and Biocomposites Engineering Applications*, 1st ed.; Wiley: Hoboken, NJ, USA, 2023; ISBN 9781119160137.
104. Liang, X.; Cha, D.K.; Xie, Q. Properties, Production, and Modification of Polyhydroxyalkanoates. *Resour. Conserv. Recycl. Adv.* **2024**, *21*, 200206. [CrossRef]
105. Tanaka, T.; Fujita, M.; Takeuchi, A.; Suzuki, Y.; Uesugi, K.; Ito, K.; Fujisawa, T.; Doi, Y.; Iwata, T. Formation of Highly Ordered Structure in Poly[(*R*)-3-Hydroxybutyrate-*co*-(*R*)-3-Hydroxyvalerate] High-Strength Fibers. *Macromolecules* **2006**, *39*, 2940–2946. [CrossRef]
106. Park, H.; He, H.; Yan, X.; Liu, X.; Scrutton, N.S.; Chen, G.-Q. PHA Is Not Just a Bioplastic! *Biotechnol. Adv.* **2024**, *71*, 108320. [CrossRef] [PubMed]
107. Pieters, K.; Mekonnen, T.H. Stable Aqueous Dispersions of Poly(3-Hydroxybutyrate-Co-3-Hydroxyvalerate) (PHBV) Polymer for Barrier Paper Coating. *Prog. Org. Coat.* **2024**, *187*, 108101. [CrossRef]
108. Dai, Z.; Li, B.; He, M.; Li, J.; Zou, X.; Wang, Y.; Shan, Y. Biodegradable and High-Performance Composites of Poly (Butylene Adipate-Co-Terephthalate) with PHBV Chain-Functionalized Nanocellulose. *Prog. Org. Coat.* **2024**, *187*, 108136. [CrossRef]
109. Jahangiri, F.; Mohanty, A.K.; Pal, A.K.; Shankar, S.; Rodriguez-Uribe, A.; Clemmer, R.; Gregori, S.; Misra, M. PHBV Coating on Biodegradable Plastic Sheet: Effect of Coating on Morphological, Mechanical and Barrier Properties. *Prog. Org. Coat.* **2024**, *189*, 108270. [CrossRef]
110. Kovalcik, A. Recent Advances in 3D Printing of Polyhydroxyalkanoates: A Review. *EuroBiotech J.* **2021**, *5*, 48–55. [CrossRef]

111. Uddin, M.d.K.; Novembre, L.; Greco, A.; Sannino, A. Polyhydroxyalkanoates, A Prospective Solution in the Textile Industry—A Review. *Polym. Degrad. Stab.* **2024**, *219*, 110619. [CrossRef]
112. You, X.; Zhou, Y.; Jin, X.; Xiang, S.; Pei, X.; Zhou, H.; Liao, Z.; Tan, Y. Green Fabrication of PHBV Microbeads Using a Dimethyl Isosorbide Solvent for Skin Exfoliators. *Green Chem.* **2023**, *25*, 9948–9958. [CrossRef]
113. Patel, R.; Gómez-Cerezo, M.N.; Huang, H.; Grøndahl, L.; Lu, M. Degradation Behaviour of Porous Poly(Hydroxybutyrate-Co-Hydroxyvalerate) (PHBV) Scaffolds in Cell Culture. *Int. J. Biol. Macromol.* **2024**, *257*, 128644. [CrossRef] [PubMed]
114. Kalva, S.N.; Dalvi, Y.B.; Khanam, N.; Varghese, R.; Ahammed, I.; Augustine, R.; Hasan, A. Air-Jet Spun PHBV/PCL Blend Tissue Engineering Scaffolds Exhibit Improved Mechanical Properties and Cell Proliferation. *Results Mater.* **2023**, *19*, 100415. [CrossRef]
115. Bayrami, S.; Chamani, M.; JamaliMoghadamSiahkali, S.; SeyedAlinaghi, S.; Shirmard, L.R.; Bayrami, S.; Javar, H.A.; Ghahremani, M.H.; Amini, M.; Tehrani, M.R.; et al. Preparation, Characterization and in Vitro Evaluation of Insulin-PHBV Nanoparticles/Alginate Hydrogel Composite System for Prolonged Delivery of Insulin. *J. Pharm. Sci.* 2024; *in press*. [CrossRef]
116. Tebaldi, M.L.; Maia, A.L.C.; Poletto, F.; De Andrade, F.V.; Soares, D.C.F. Poly(-3-Hydroxybutyrate-Co-3-Hydroxyvalerate) (PHBV): Current Advances in Synthesis Methodologies, Antitumor Applications and Biocompatibility. *J. Drug Deliv. Sci. Technol.* **2019**, *51*, 115–126. [CrossRef]
117. Meng, D.-C.; Shen, R.; Yao, H.; Chen, J.-C.; Wu, Q.; Chen, G.-Q. Engineering the Diversity of Polyesters. *Curr. Opin. Biotechnol.* **2014**, *29*, 24–33. [CrossRef]
118. Rabello, L.G.; Ribeiro, R.C.D.C.; Pinto, J.C.C.D.S.; Thiré, R.M.D.S.M. Chemical Recycling of Green Poly(3-Hydroxybutyrate-Co-3-Hydroxyvalerate) (PHBV)-Based Air Filters through Hydrolysis. *J. Environ. Chem. Eng.* **2024**, *12*, 111816. [CrossRef]
119. Lyshtva, P.; Voronova, V.; Barbir, J.; Leal Filho, W.; Kröger, S.D.; Witt, G.; Miksch, L.; Saborowski, R.; Gutow, L.; Frank, C.; et al. Degradation of a Poly(3-Hydroxybutyrate-Co-3-Hydroxyvalerate) (PHBV) Compound in Different Environments. *Heliyon* **2024**, *10*, e24770. [CrossRef] [PubMed]
120. Ho, Y.-H.; Gan, S.-N.; Tan, I.K.P. Biodegradation of a Medium-Chain-Length Polyhydroxyalkanoate in Tropical River Water. *ABAB* **2002**, *102–103*, 337–348. [CrossRef] [PubMed]
121. Arcos-Hernandez, M.V.; Laycock, B.; Pratt, S.; Donose, B.C.; Nikolić, M.A.L.; Luckman, P.; Werker, A.; Lant, P.A. Biodegradation in a Soil Environment of Activated Sludge Derived Polyhydroxyalkanoate (PHBV). *Polym. Degrad. Stab.* **2012**, *97*, 2301–2312. [CrossRef]
122. Weng, Y.-X.; Wang, X.-L.; Wang, Y.-Z. Biodegradation Behavior of PHAs with Different Chemical Structures under Controlled Composting Conditions. *Polym. Test.* **2011**, *30*, 372–380. [CrossRef]
123. Salomez, M.; George, M.; Fabre, P.; Touchaleaume, F.; Cesar, G.; Lajarrige, A.; Gastaldi, E. A Comparative Study of Degradation Mechanisms of PHBV and PBSA under Laboratory-Scale Composting Conditions. *Polym. Degrad. Stab.* **2019**, *167*, 102–113. [CrossRef]
124. Shah, A.A.; Hasan, F.; Hameed, A. Degradation of Poly(3-Hydroxybutyrate-Co-3-Hydroxyvalerate) by a Newly Isolated *Actinomadura* Sp. AF-555, from Soil. *Int. Biodeterior. Biodegrad.* **2010**, *64*, 281–285. [CrossRef]
125. Calabia, B.P.; Tokiwa, Y. A Novel PHB Depolymerase from a Thermophilic *Streptomyces* sp. *Biotechnol. Lett.* **2006**, *28*, 383–388. [CrossRef] [PubMed]
126. Kolter, K.; Dashevsky, A.; Irfan, M.; Bodmeier, R. Polyvinyl acetate-based film coatings. *Int J Pharm.* **2013**, *457*, 470–479. [CrossRef] [PubMed]
127. Rathi, V.; Brajpuriya, R.; Gupta, R.; Parmar, K.P.S.; Kumar, A. Graphene-Derived Composites: A New Frontier in Thermoelectric Energy Conversion. *Energy Adv.* **2024**, *3*, 389–412. [CrossRef]
128. Amann, M.; Minge, O. Biodegradability of Poly(Vinyl Acetate) and Related Polymers. In *Synthetic Biodegradable Polymers*; Rieger, B., Künkel, A., Coates, G.W., Reichardt, R., Dinjus, E., Zevaco, T.A., Eds.; Springer: Berlin/Heidelberg, Germany, 2011; Volume 245, pp. 137–172. ISBN 9783642271533.
129. Kricheldorf, H.R.; Nuyken, O.; Swift, G. (Eds.) *Handbook of Polymer Synthesis: Second Edition*; CRC Press: Boca Raton, FL, USA, 2004; ISBN 9780429117008.
130. Cappitelli, F.; Sorlini, C. Microorganisms Attack Synthetic Polymers in Items Representing Our Cultural Heritage. *Appl. Environ. Microbiol.* **2008**, *74*, 564–569. [CrossRef] [PubMed]
131. Liu, Z.; Li, G.; Zhang, F.; Wu, J. Enhanced Biodegradation Activity towards Poly(Ethyl Acrylate) and Poly(Vinyl Acetate) by Anchor Peptide Assistant Targeting. *J. Biotechnol.* **2022**, *349*, 47–52. [CrossRef] [PubMed]
132. Dhanasekaran, N.P.D.; Muthuvelu, K.S.; Arumugasamy, S.K. Recent Advancement in Biomedical Applications of Polycaprolactone and Polycaprolactone-Based Materials. In *Encyclopedia of Materials: Plastics and Polymers*; Elsevier: Amsterdam, The Netherlands, 2022; pp. 795–809. ISBN 9780128232910.
133. Patra, S.; Singh, M.; Pareek, D.; Wasnik, K.; Gupta, P.S.; Paik, P. Advances in the Development of Biodegradable Polymeric Materials for Biomedical Applications. In *Encyclopedia of Materials: Plastics and Polymers*; Elsevier: Amsterdam, The Netherlands, 2022; pp. 532–566. ISBN 9780128232910.
134. Kumar, S.; Singh, R. Bio-Compatible Polymer Matrix for 3D Printing: A Review. In *Encyclopedia of Materials: Plastics and Polymers*; Elsevier: Amsterdam, The Netherlands, 2022; pp. 39–46, ISBN 9780128232910.
135. Gupta, D.; Dogra, V.; Verma, D.; Chaudhary, A.K.; Tewari, M. PCL-Based Composites and Their Utilizations in the Medical Sector. In *Bioresorbable Polymers and Their Composites*; Elsevier: Amsterdam, The Netherlands, 2024; pp. 63–83, ISBN 9780443189159.

136. Hou, Y.; Wang, W.; Bartolo, P. Investigation of Polycaprolactone for Bone Tissue Engineering Scaffolds: In Vitro Degradation and Biological Studies. *Mater. Des.* **2022**, *216*, 110582. [CrossRef]
137. Yadav, A.L.; Gurave, P.M.; Gadkari, R.R.; Wazed Ali, S. PCL-Based Bionanocomposites in Tissue Engineering and Regenerative Medicine. In *Bionanocomposites in Tissue Engineering and Regenerative Medicine*; Elsevier: Amsterdam, The Netherlands, 2021; pp. 465–480, ISBN 9780128212806.
138. Wypych, G. PCL Poly(ε-Caprolactone). In *Handbook of Polymers*; Elsevier: Amsterdam, The Netherlands, 2016; pp. 323–325, ISBN 9781895198928.
139. Chen, D.R.; Bei, J.Z.; Wang, S.G. Polycaprolactone Microparticles and Their Biodegradation. *Polym. Degrad. Stab.* **2000**, *67*, 455–459. [CrossRef]
140. Richert, A.; Dąbrowska, G.B. Enzymatic Degradation and Biofilm Formation during Biodegradation of Polylactide and Polycaprolactone Polymers in Various Environments. *Int. J. Biol. Macromol.* **2021**, *176*, 226–232. [CrossRef]
141. Yagi, H.; Ninomiya, F.; Funabashi, M.; Kunioka, M. Anaerobic Biodegradation Tests of Poly(Lactic Acid) and Polycaprolactone Using New Evaluation System for Methane Fermentation in Anaerobic Sludge. *Polym. Degrad. Stab.* **2009**, *94*, 1397–1404. [CrossRef]
142. Al Hosni, A.S.; Pittman, J.K.; Robson, G.D. Microbial Degradation of Four Biodegradable Polymers in Soil and Compost Demonstrating Polycaprolactone as an Ideal Compostable Plastic. *Waste Manag.* **2019**, *97*, 105–114. [CrossRef] [PubMed]
143. Lefebvre, F.; David, C.; Vander Wauven, C. Biodegradation of Polycaprolactone by Micro-Organisms from an Industrial Compost of Household Refuse. *Polym. Degrad. Stab.* **1994**, *45*, 347–353. [CrossRef]
144. Eling, B.; Tomović, Ž.; Schädler, V. Current and Future Trends in Polyurethanes: An Industrial Perspective. *Macro Chem. Phys.* **2020**, *221*, 2000114. [CrossRef]
145. Santos-Beneit, F.; Chen, L.M.; Bordel, S.; Frutos de la Flor, R.; García-Depraect, O.; Lebrero, R.; Rodriguez-Vega, S.; Muñoz, R.; Börner, R.A.; Börner, T. Screening Enzymes That Can Depolymerize Commercial Biodegradable Polymers: Heterologous Expression of *Fusarium solani* Cutinase in Escherichia coli. *Microorganisms* **2023**, *11*, 328. [CrossRef]
146. Garrido, S.M.; Kitamoto, N.; Watanabe, A.; Shintani, T.; Gomi, K. Functional Analysis of FarA Transcription Factor in the Regulation of the Genes Encoding Lipolytic Enzymes and Hydrophobic Surface Binding Protein for the Degradation of Biodegradable Plastics in *Aspergillus oryzae*. *J. Biosci. Bioeng.* **2012**, *113*, 549–555. [CrossRef] [PubMed]

International Journal of
Molecular Sciences

MDPI

Article

Use of 3-Deoxy-D-arabino-heptulosonic acid 7-phosphate Synthase (DAHP Synthase) to Enhance the Heterologous Biosynthesis of Diosmetin and Chrysoeriol in an Engineered Strain of *Streptomyces albidoflavus*

Álvaro Pérez-Valero [1,2,3], Juan Serna-Diestro [1,2,3], Claudio J. Villar [1,2,3] and Felipe Lombó [1,2,3,*]

1 Research Group BIONUC (Biotechnology of Nutraceuticals and Bioactive Compounds), Departamento de Biología Funcional, Área de Microbiología, Universidad de Oviedo, 33006 Oviedo, Spain; apv.moratalla@gmail.com (Á.P.-V.); uo264212@uniovi.es (J.S.-D.); cjvg@uniovi.es (C.J.V.)

2 IUOPA (Instituto Universitario de Oncología del Principado de Asturias), 33006 Oviedo, Spain

3 ISPA (Instituto de Investigación Sanitaria del Principado de Asturias), 33011 Oviedo, Spain

* Correspondence: lombofelipe@uniovi.es

Abstract: Flavonoids are a large family of polyphenolic compounds with important agro-industrial, nutraceutical, and pharmaceutical applications. Among the structural diversity found in the flavonoid family, methylated flavonoids show interesting characteristics such as greater stability and improved oral bioavailability. This work is focused on the reconstruction of the entire biosynthetic pathway of the methylated flavones diosmetin and chrysoeriol in *Streptomyces albidoflavus*. A total of eight different genes (TAL, 4CL, CHS, CHI, FNS1, F3′H/CPR, 3′-OMT, 4′-OMT) are necessary for the heterologous biosynthesis of these two flavonoids, and all of them have been integrated along the chromosome of the bacterial host. The biosynthesis of diosmetin and chrysoeriol has been achieved, reaching titers of 2.44 mg/L and 2.34 mg/L, respectively. Furthermore, an additional compound, putatively identified as luteolin 3′,4′-dimethyl ether, was produced in both diosmetin and chrysoeriol-producing strains. With the purpose of increasing flavonoid titers, a 3-Deoxy-D-arabino-heptulosonic acid 7-phosphate synthase (DAHP synthase) from an antibiotic biosynthetic gene cluster (BGC) from *Amycolatopsis balhimycina* was heterologously expressed in *S. albidoflavus*, enhancing diosmetin and chrysoeriol production titers of 4.03 mg/L and 3.13 mg/L, which is an increase of 65% and 34%, respectively. To the best of our knowledge, this is the first report on the de novo biosynthesis of diosmetin and chrysoeriol in a heterologous host.

Keywords: flavonoid; biosynthesis; *O*-methyltransferase; regiospecificity; metabolic engineering

1. Introduction

Flavonoids are a large class of polyphenols, representing around 9000 compounds that are widely distributed in plants [1–3]. These bioactive compounds are important in the nutraceutical, pharmaceutical, cosmetic, and agro-industrial fields due to the vast variety of properties they display [4–6], such as antitumor [7–9], antimicrobial, antiangiogenic [8,9], antioxidant, and neuroprotective compounds, among other bioactivities [10]. Flavonoids undergo diverse enzymatic modifications in their core structure, such as hydroxylations, glycosylations, or methylations, and in particular, methylated flavonoids possess interesting properties, such as major stability, improved oral bioavailability, enhanced membrane transport, and better intestinal absorption [11]. In the case of the methylated flavone diosmetin, which is widely present (as a glucosylated derivative) in several *Citrus* fruits, it has been reported to have antioxidant, phlebotonic, oestrogenic, antimicrobial, and anti-inflammatory activities [12]. As an antimicrobial, it should be noted that in combination with erythromycin, diosmetin shows a synergistic effect against methicillin-resistant *Staphylococcus aureus* (MRSA) [13]. On the other hand, chrysoeriol, another luteolin methylated derivative, has been found in several plants, such as *Reseda luteola*, *Melientha suavis*, and *Cardiospermum halicacabum* L. This methylated flavonoid is of interest due to its potential as an anti-hyperlipidemic, antitumor, antioxidant, antimicrobial, antifungal, and neuroprotective agent, among other bioactivities [14].

Depending on their structural differences, flavonoids are classified into seven subclasses: flavanones, flavones, isoflavones, flavonols, anthocyanidins, flavanols, and chalcones [15]. For the heterologous biosynthesis of flavonoids in bacteria, several enzymatic steps must take place, starting with the conversion of L-tyrosine in coumaric acid by the tyrosine ammonia lyase (TAL). Then, coumaric acid should be activated with the molecule coenzyme-A by the action of 4-coumaroyl-CoA ligase (4CL), giving rise to 4-coumaroyl-CoA, which would be then condensed with three molecules of malonyl-CoA through chalcone synthase (CHS), generating naringenin chalcone, the basic carbon skeleton for all flavonoids known in nature [16–19]. Finally, the heterocycle closure in naringenin chalcone is carried out by a chalcone isomerase (CHI), generating the universal flavanone precursor, naringenin. For the biosynthesis of diosmetin and chrysoeriol, three extra enzymatic steps are necessary. Naringenin should be converted to apigenin by a flavone synthase. In the bacterial host *S. albidoflavus*, a class I flavone synthase (FNS1) is necessary, lacking the need for a cytochrome P450 membrane-bound monooxygenase, which would be difficult to express in bacteria [20]. Apigenin should then be converted to luteolin by a flavone 3′ hydroxylase. In this work, a flavone 3′ hydroxylase coupled with a soluble cytochrome P450 reductase (F3′H-CPR) has been used, providing the reducing power to the flavone 3′ hydroxylase [21]. Finally, a 4′-*O*-methyltransferase (4′OMT) is necessary to reach the heterologous biosynthesis of diosmetin, and a 3′-*O*-methyltransferase (3′OMT) is required for the heterologous production of chrysoeriol.

The biosynthesis of methylated flavonoids has been achieved using different microbial cell factories, such as *Escherichia coli* [22] and *Saccharomyces cerevisiae* [23]. Additionally, in recent work by our research group, the biosynthesis of this type of flavonoid has been achieved in the Gram-positive bacterium *S. albidoflavus* [24]. However, a major bottleneck in the heterologous biosynthesis of flavonoids in bacteria using synthetic biology tools is the low production titters. Several strategies have been applied to increase the intracellular pools of flavonoid precursors, such as the redirection of the carbon source for the bacteria towards the biosynthesis of malonyl-CoA [25], a key precursor in the flavonoid pathway.

So far, the biosynthesis of diosmetin and chrysoeriol had never been carried out in a heterologous host. In this work, we present the production of both methylated flavones after dissecting their biosynthetic pathways into three modules, leveraging different integration sites of the bacteriophages ϕC31 [26] and ϕBT1 [27], in addition to the integration site of the pSAM2 plasmid [28] in the chromosome of *S. albidoflavus*, enabling the stable incorporation of exogenous DNA into the bacterial chromosome (across three different sections, corresponding to three different steps during flavonoid biosynthesis) and avoiding the necessity for antibiotic selection to maintain plasmids.

As a further step towards the enhancement of the final production titers of these two methylated flavones in *S. albidoflavus*, the gene encoding a DAHP synthase from the actinomycete *Amycolatopsis balhimycina* [29] has been added together with the diosmetin and chrysoeriol biosynthetic pathways. This enzyme, DAHP, carries out the condensation of phosphoenolpyruvate (PEP) and erythrose 4-phosphate (E4P) [30], the first enzymatic reaction in the shikimate pathway towards the biosynthesis of aromatic amino acids, such as the flavonoid precursor L-tyrosine.

The current industrial production of diosmetin is based on semi-synthesis approaches from its glycosylated form, diosmin. This semi-synthesis makes use of concentrated sulfuric acid and several crystallization steps, which lower the final efficiency of the whole process. Other industrial alternatives include the use of hesperidin (a flavanone from orange peels), which is oxidized under hot alcoholic sodium acetate (with its sugar moiety removed chemically) to generate diosmetin [31,32]. In this work, the need for flavonoid precursors (hesperidin, diosmin, etc.) is not necessary for obtaining the final bioactive compounds, as these are generated de novo, using common metabolic intermediates available in the bacterial cytoplasm.

2. Results

2.1. Heterologous Biosynthesis of Diosmetin

For the biosynthesis of diosmetin in bacteria, the action of seven enzymes (TAL, 4CL, CHS, CHI, FNS1, F3′H-CPR, and 4′OMT) is required (in contrast to the eight necessary genes in plants). In this work, this whole biosynthetic pathway was divided into three modules (Figure 1), and the necessary coding genes were distributed along the chromosome of the strain *S. albidoflavus* UO-FLAV-004 [24], taking advantage of the different prophage integration sites in this species. All the genes were codon-optimized for *S. albidoflavus*. TAL, 4CL, CHS, and CHI genes were integrated into the ϕC31 *attb* site [24], FNS1 [24] and F3′H-CPR [21] genes were integrated into the ϕBT1 *attb* site, and the 4′OMT gene was integrated into the pSAM2 site of the chromosome.

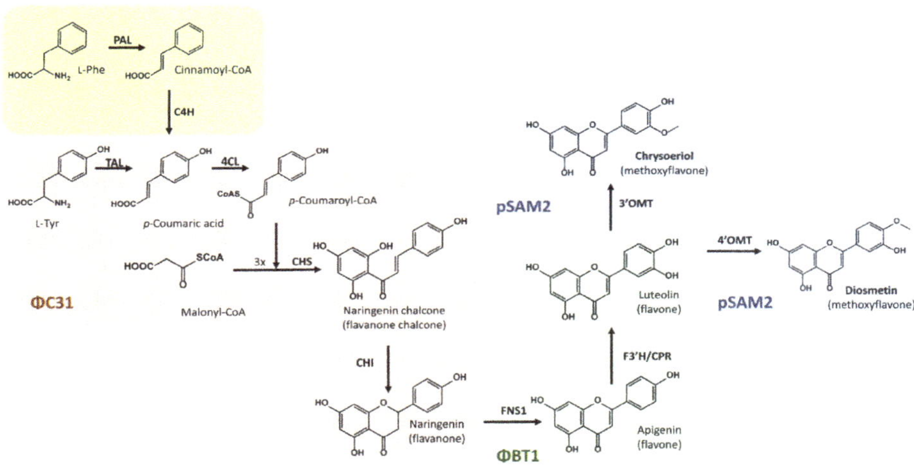

Figure 1. Biosynthetic pathway for the biosynthesis of diosmetin and chrysoeriol. The yellow section indicates the canonical first steps in plants. Tyrosine ammonia-lyase (TAL); 4-Coumaroyl-CoA ligase (4CL); Chalcone synthase (CHS); Chalcone isomerase (CHI); Flavone synthase (FNS1); Flavonoid 3′ hydroxylase/Cytochrome P450 reductase chimera (F3′H/CPR); 4′-*O*-methyltransferase (4′OMT); 3′-*O*-methyltransferase (3′OMT). In light orange, the part of the pathway integrated into the ϕC31 *attb* site of the chromosome is shown; in light green, the part of the pathway integrated into the ϕBT1 *attb* site of the chromosome is shown; in light blue, the final part of the pathway for the heterologous biosynthesis of diosmetin or chrysoeriol is shown, with both cases integrated into the pSAM2 site of the *S. albidoflavus* chromosome. The yellow section indicates the naturally occurring pathway in plants, using L-phenylalanine instead of L-tyrosine.

The selected 4′OMT is a hypothetical class I SAM-dependent *O*-methyltransferase (M444_29925) from *Streptomyces* sp. Mg1 (see Section 4). This gene was assembled under the control of the SP25 promoter, which shows good transcriptional activity, being higher than the widely used PermE* promoter [33]. The gene *M444_29925* was selected after a BLASTP analysis against the *O*-methyltransferase GerMIII, which shows a relative in vitro conversion rate of luteolin to diosmetin of 67% [34]. The differences between *GerMIII* and *M444_29925* at the protein level lie in the amino acids 81 (A-P) and 377 (T-I).

After the corresponding fermentation experiments, the control strain *S. albidoflavus* UO-FLAV-004-LUT, which harbors the genes TAL, 4CL, CHS, CHI, FNS1, and F3′H-CPR, as well as an empty plasmid integrated into the pSAM2 chromosomal site, was able to produce the unmethylated precursor of diosmetin and luteolin, reaching titers of 4.61 mg/L (Figure 2A). The diosmetin-producing strain *S. albidoflavus* UO-FLAV-004-DIO, which additionally containins the gene coding for the M444_29925 4′-*O*-methyltransferase integrated into the pSAM2 chromosomal site, generated 2.44 mg/L of diosmetin and 0.73 mg/L of luteolin. Surprisingly, a small peak of chrysoeriol was also detected in the chromatogram, indicating the capability of this strain to produce the luteolin 3′-*O*-methyl ether (chrysoeriol) at titers of 0.38 mg/L as well (Figure 2B).

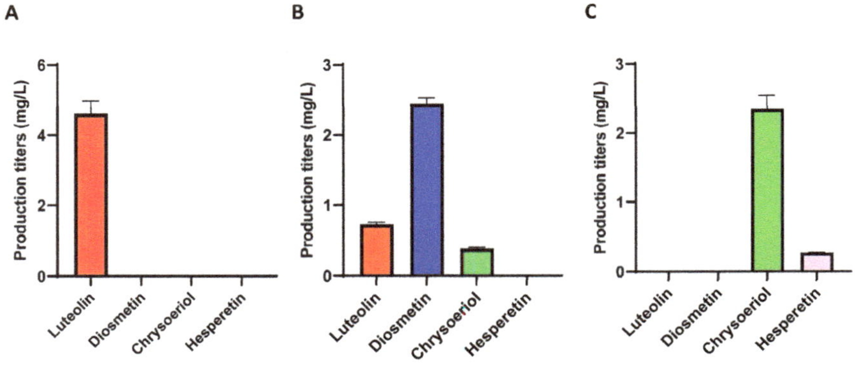

Figure 2. Production titers of different flavonoids in different strains derived from *S. albidoflavus* UO-FLAV-004. (**A**) *S. albidoflavus* UO-FLAV-004-LUT; (**B**) *S. albidoflavus* UO-FLAV-004-DIO; (**C**) *S. albidoflavus* UO-FLAV-004-CHR.

2.2. Heterologous Biosynthesis of Chrysoeriol

The heterologous biosynthesis of chrysoeriol was carried out using the same distribution of genes as in the previous case. The same enzymatic activities required for the heterologous biosynthesis of diosmetin were necessary to produce the common precursor, luteolin, but a different *O*-methyltransferase was needed to introduce the methyl moiety at position 3′ instead of position 4′ of the ring B of luteolin (Figure 1). The 3′-*O*-methyltransferase used in the biosynthesis of chrysoeriol was a CCoAOMT-like enzyme from *Arabidopsis thaliana* (At4g26220), whose gene was optimized for *S. albidoflavus* (Table 1). This enzyme possessed a greater preference for introducing a methyl group in the *para* position (4′) in flavanones and dihydroflavonols, whereas flavones and flavonols were methylated in the meta position (3′) [35]. Using this enzyme, the biosynthesis of chrysoeriol in the strain *S. albidoflavus* UO-FLAV-004-CHR was achieved, reaching titers of 2.34 mg/L, with no remaining luteolin detected. Also, a small peak corresponding to hesperetin was detected in the chromatogram of this culture extract, and its production reached 0.27 mg/L (Figure 2C). The presence of this compound in this extract will be addressed in the Section 3. The strain used as control for this fermentation was *S. albidoflavus* UO-FLAV-004-LUT, as in the previous case.

2.3. Identification of Putative Luteolin 3′,4′-Dimethyl Ether in Both Diosmetin- and Chrysoeriol-Producing Strains

After analyzing the chromatograms of the two different producing strains, an extra differential peak was also detected at late retention times (35.4 min), with a higher intensity in the case of the strain *S. albidoflavus* UO-FLAV-004-DIO (Figure 3). A late retention time indicates less polarity in the used HPLC program (see Section 4), which suggested that this peak could correspond to di-methylated luteolin, a less polar compound than a single methylated luteolin. The available positions in diosmetin for *O*-methylation were positions 5 and 7 of ring A and position 3′ of ring B, while for chrysoeriol, the available positions were positions 5 and 7 of ring A and position 4′ of ring B. Since standards for all possible combinations were not available, a feeding assay was performed to confirm or discard our first putative candidate, which putatively was considered luteolin 3′,4′-dimethyl ether.

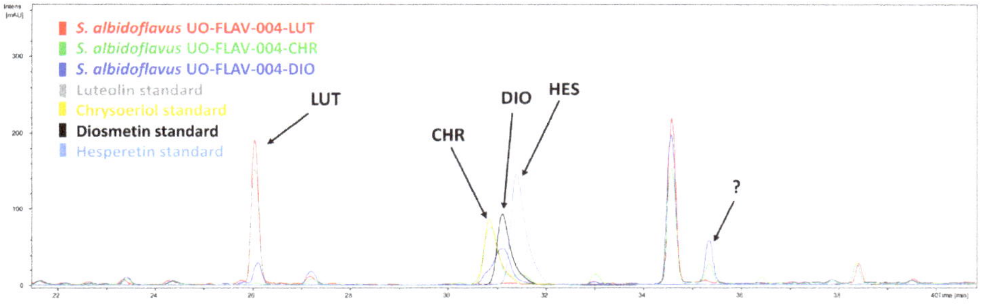

Figure 3. HPLC-DAD chromatograms of the strains *S. albidoflavus* UO-FLAV-004-LUT (red), *S. albidoflavus* UO-FLAV-004-CHR (green), *S. albidoflavus* UO-FLAV-004-DIO (blue). The four commercial standards are also indicated. Luteolin (LUT); Chrysoeriol (CHR); Diosmetin (DIO); Hesperetin (HES); Unknown differential peak (?).

In the strains *S. albidoflavus* UO-FLAV-004-DIO and *S. albidoflavus* UO-FLAV-004-CHR, one of the enzymes in the heterologous biosynthetic pathway was the FNS1 flavone synthase from *Petroselinum crispum*, which has been described as being able to transform homoeriodictyol to chrysoeriol with high efficiency [36] (Figure 4). Taking this into account, since the strain *S. albidoflavus* UO-FLAV-004-CHR harbored the necessary genes for the biosynthesis of hesperetin (eriodictyol 4′-*O*-methyl ether), such as the *O*-methyltransferase At4g26220 [35] (Figure 4), hesperetin could be converted to homohesperetin (eriodictyol 3′,4′-dimethyl ether) by an endogenous *O*-methyltransferase activity of *S. albidoflavus* (see Section 3). Due to the structural similarity between homoeriodictyol (eriodictyol 3′-*O*-methyl ether) and homohesperetin (eriodictyol 3′,4′-*O*-methyl ether), we hypothesize that the FNS1 enzyme could be acting on the homohesperetin flavanone to produce the luteolin 3′,4′-dimethyl ether flavone (Figure 4).

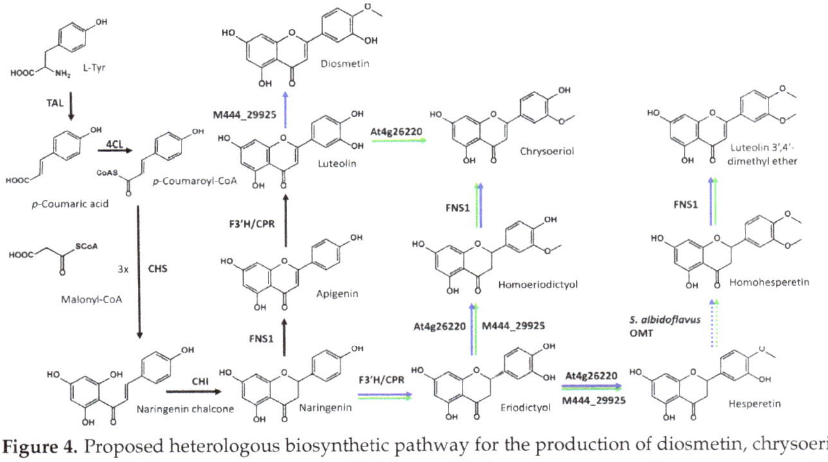

Figure 4. Proposed heterologous biosynthetic pathway for the production of diosmetin, chrysoeriol, and luteolin 3′,4′-dimethyl ether in *S. albidoflavus*. Black arrows connect the necessary reactions to reach the biosynthesis of luteolin. Blue arrows represent the reactions carried out in the strain *S. albidoflavus* UO-FLAV-004-DIO. Finally, green arrows represent the reactions carried out in the strain *S. albidoflavus* UO-FLAV-004-CHR. Dashed arrows indicate a reaction carried out by an endogenous enzyme of *S. albidoflavus*.

With the aim of proving the presence of this extra activity in FNS1, a feeding was made with homohesperetin at a final concentration of 0.1 mM to the strain *S. albidoflavus* UO-FLAV-004-FNS1, previously developed by our research group [24], and to the strain *S. albidoflavus* UO-FLAV-004 as a control. Analysis of the HPLC-DAD chromatograms generated showed the presence of a derivative peak in the strain containing FNS1, which was co-eluting with the extra peaks observed in the diosmetin- and chrysoeriol-producing strains. This new peak was also showed the same UV absorption spectrum as the putative luteolin 3′,4′-dimethyl ether from the diosmetin- and chrysoeriol-producing strains (Figure S1). These results suggested that the strain *S. albidoflavus* UO-FLAV-004-CHR was able to produce luteolin 3′,4′-dimethyl ether through this pathway (Figure 4), and it indicated that homohesperetin could be a good substrate for the FNS1 enzyme.

On the other hand, the strain *S. albidoflavus* UO-FLAV-004-DIO, which also produced this putative luteolin di-methylated derivative, albeit at a slightly higher concentration, should not be able to produce it through the same pathway as *S. albidoflavus* UO-FLAV-004-CHR, since the gene *At4g26220* was not present in this strain. A feasible alternative to explain the production of this compound in this strain was the conversion of eriodictyol to hesperetin by the action of the enzyme 4′OMT of *Streptomyces* sp. Mg1, an activity that had not been reported so far. To check the substrate flexibility of this enzyme, a feeding experiment using eriodictyol was performed on the strain *S. albidoflavus* UO-FLAV-004-M444_29925, harboring only the 4′OMT of *Streptomyces* sp. Mg1 in the pSAM2 chromosomal integration site, along with the corresponding control strain *S. albidoflavus* UO-FLAV-004 harboring an empty plasmid in the same chromosomal *attb* site. The feeding results were analyzed by HPLC-DAD and showed a good conversion of eriodictyol to both hesperetin (eriodictyol 4′-*O*-methyl ether) and homoeriodictyol (eriodictyol 3′-*O*-methyl ether) in *S. albidoflavus* UO-FLAV-004-M444_29925. Additionally, a peak of homohesperetin was detected after the feeding (Figure S2). No eriodictyol derivative was observed in the control strain. This suggested that hesperetin was generated in the strain *S. albidoflavus* UO-FLAV-004-DIO by the enzyme M444_29925 and converted to homohesperetin by the putative endogenous *O*-methyltransferase activity of *S. albidoflavus*, finally being converted to luteolin 3′,4′-dimethyl ether by the action of FNS1, like in the case observed in the strain *S. albidoflavus* UO-FLAV-004-CHR (Figure 4).

2.4. Use of a DAHP Synthase to Increase the Production Titers of Diosmetin and Chrysoeriol through Precursor Titer Enhancement

DAHP synthase is the first enzyme of the shikimate pathway. This enzyme condenses D-erythrose 4-phosphate and phosphoenolpyruvate to produce DAHP, a key precursor in the biosynthesis of aromatic amino acids, such as L-tyrosine (Figure 5). L-tyrosine is the first precursor in the heterologous pathway for flavonoid biosynthesis in *S. albidoflavus*, making the overexpression of the DAHP synthase an interesting strategy to enhance the intracellular pools of L-tyrosine and, thus, the final flavonoid titers. This strategy was followed by Thykaer and colleagues to increase the titers of the vancomycin analogue balhimycin in the natural producing strain of *Amycolatopsis balhimycina*, resulting in improved specific productivities of balhimycin by introducing an extra copy of the *dahp* gene [29].

In this work, a codon optimization of the *dahp* gene of *Amycolatopsis balhimycina* was conducted for *S. albidoflavus*, and the gene was assembled under the control of the strong SP25 promoter. Then, this gene was brought together with the different *O*-methyltransferases involved in diosmetin and chrysoeriol biosynthesis (see Section 4) and integrated into the pSAM2 chromosomal site of *S. albidoflavus* UO-FLAV-004-LUT parental strain, giving rise to the final strains *S. albidoflavus* UO-FLAV-004-DIO-dahp and *S. albidoflavus* UO-FLAV-004-CHR-dahp, respectively. These new strains were cultivated at the same time as the strains *S. albidoflavus* UO-FLAV-004-DIO and *S. albidoflavus* UO-FLAV-004-CHR, used as controls, to check if the enzyme DAHP synthase was able to increase the titers of diosmetin and chrysoeriol. The resulting production titers were 4.03 mg/L of diosmetin for *S. albidoflavus* UO-FLAV-004-DIO-dahp and 3.13 mg/L of chrysoeriol for *S. albidoflavus* UO-FLAV-004-CHR-dahp, which represents a 1.65-fold and 1.34-fold increase, respectively (Figure 6). Also, as expected, the by-products chrysoeriol and hesperetin were present in the strains *S. albidoflavus* UO-FLAV-004-DIO and *S. albidoflavus* UO-FLAV-004-CHR, respectively, as well as the putative luteolin 3′,4′-dimethyl ether, all in proportionally increased quantities in the strains *S. albidoflavus* UO-FLAV-004-DIO-dahp and *S. albidoflavus* UO-FLAV-004-CHR-dahp (Figures S3 and S4).

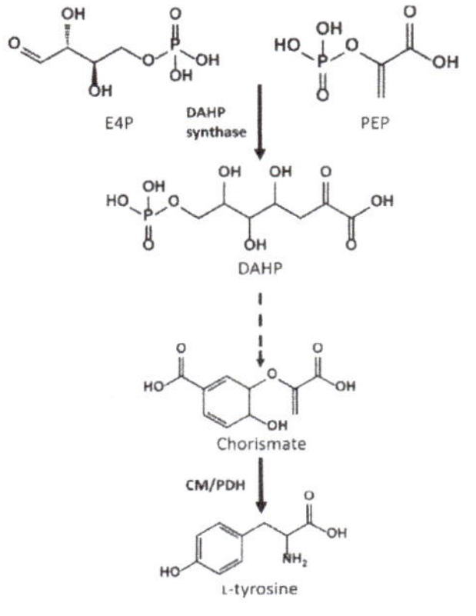

Figure 5. Abbreviated schema of the shikimate pathway for the generation of L-tyrosine. E4P: Erythrose 4-phosphate; PEP: phosphoenolpyruvate; DAHP: 3-Deoxy-D-arabino-heptulosonic acid 7-phosphate: CM/PDH: chorismate mutase/prephenate dehydrogenase.

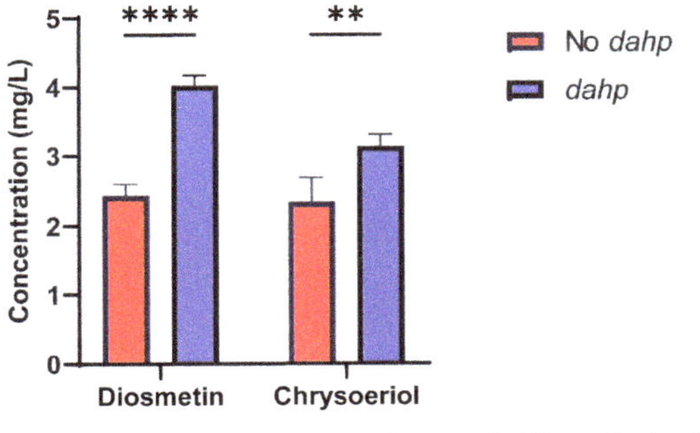

Figure 6. Effect of the DAHP synthase on the biosynthesis of diosmetin and chrysoeriol in the strains *S. albidoflavus* UO-FLAV-004-DIO-dahp and *S. albidoflavus* UO-FLAV-004-CHR-dahp compared to the strains *S. albidoflavus* UO-FLAV-004-DIO and *S. albidoflavus* UO-FLAV-004-CHR, respectively. Asterisks indicate statistically significant differences (** $p < 0.005$; **** $p < 0.0001$).

3. Discussion

Previous studies conducted by our research group have revealed the potential of *S. albidoflavus* for the biosynthesis of methylated derivatives of flavonoids [24]. To our knowledge, this work describes, for the first time, the complete biosynthesis of the methylated flavonoids diosmetin and chrysoeriol in a heterologous host.

The enzyme M444_29925 shares 98% identity with the well-studied GermIII *O*-methyltransferase, which shows high regiospecificity for the 4′ position of flavones [34]. Along the biosynthetic pathway of diosmetin, the reactions can also occur in a different order. If the chimeric F3′H-CPR hydroxylase uses naringenin as a substrate before FNS1, the strain will first produce eriodictyol. A difference between flavones and flavanones is the spatial configuration of their chemical structures. Flavones present a planar structure [37], while flavanones present a chair conformational structure [38]. According to this, we hypothesize that the enzyme M444_29925 may introduce a methyl group in a different position in a flavanone structure, such as eriodictyol, instead of the 4′ position, as observed in flavones. With the aim of finding out if *O*-methylation in a different position is possible for this enzyme, a feeding experiment with eriodictyol was performed on a strain containing only the *O*-methyltransferase M444_29925, named *S. albidoflavus* UO-FLAV-004-M444_29925, and the strain *S. albidoflavus* UO-FLAV-004 as a control. As shown previously, we verified that this enzyme was able to methylate eriodictyol, at least, at positions 4′ and 3′, yielding hesperetin (eriodictyol 4′-*O*-methyl ether) and homoeriodictyol (eriodictyol 3′-*O*-methyl ether) (Figure 4). The generation of homoeriodictyol through this enzyme can explain the presence of chrysoeriol in the strain *S. albidoflavus* UO-FLAV-004-DIO due to the action of FNS1 (Figure 4) [36].

On the other hand, the presence of hesperetin in the chrysoeriol-producing strain is easy to explain since the CcoAOMT-like enzyme At4g26220 efficiently produces hesperetin from eriodictyol in vitro (Figure 4) [35]. At4g26220 also produces homoeriodictyol [35], but again, FNS1 converts it to chrysoeriol.

Finally, as described in the Section 2, homohesperetin was detected after the administration of eriodictyol to the strain *S. albidoflavus* UO-FLAV-004-M444_29925. Given that luteolin 3′,4′-dimethyl ether is a direct derivative of homohesperetin following enzymatic mediation by FNS1, and this proposed compound is discernible in both *S. albidoflavus* UO-FLAV-004-DIO and *S. albidoflavus* UO-FLAV-004-CHR strains, it is inferred that homohesperetin should be inherently present in both the diosmetin- and chrysoeriol-producing strains. This deduction is supported by the detection of the putative luteolin 3′,4′-dimethyl ether in both strains. Consequently, these results imply that homohesperetin is likely produced through an unknown endogenous activity of *S. albidoflavus* in the presence of hesperetin and/or homoeriodictyol.

Thus, if homohesperetin is found in the bacterial cytoplasm at any moment during the biosynthesis of diosmetin and chrysoeriol, it can be transformed into luteolin 3′,4′-dimethyl ether by the action of the FNS1. These results highlight the important role that substrate specificity of the selected enzymes plays in the heterologous biosynthesis of natural products, and it may be the case that these enzymes compete for different molecules found at different stages of a given pathway, generating different derivatives. However, this could also be advantageous since new unexpected products could be obtained, such as the putative luteolin 3′,4′-dimethyl ether, establishing new pathways to produce them.

Regarding the biosynthesis of diosmetin and chrysoeriol, attempts were made to increase production titers by using a DAHP synthase. The biosynthesis of aromatic amino acids is strictly regulated by feedback inhibition mechanisms, and DAHP synthases are normally feedback regulated by L-tyrosine, L-phenylalanine, or both [39]. The strategy of using a DAHP synthase to enhance the flavonoid titters has been carried out by different researchers. Koopman and colleagues achieved an increase in naringenin titters using different approaches in *Saccharomyces cerevisiae*, including the alleviation of feedback inhibition of yeast DAHP synthase by introducing a L-tyrosine insensitive ARO4 (*ARO4*^{G226S}) allele in conjunction with the deletion of the other allele of the *dahp* gene [40]. In another experiment in *E. coli*, the overexpression of a feedback-resistant derivative of *dahp* (*aroG*^{fbr}), together with the overexpression of chorismate mutase/prephenate dehydrogenase (*tyrA*^{fbr}), led to an increase in naringenin [41,42], apigenin, and genkwanin titters [43].

Here, a DAHP synthase from the actinomycete *Amycolatopsis balhimycina* has been used, which has been previously proved as a useful metabolic engineering strategy to increase the biosynthesis of glycopeptide antibiotics by introducing an extra copy in its natural producer. In *A. balhimycina*, the gene encoding this DAHP synthase is found in a chromosomal region containing a BGC and is not involved in primary metabolism [29]. Other *dahp* genes have been identified in other BGCs [44,45], and in these cases, the *dahp* genes were like those encoding the plant type DAHP synthases, which were proven to be naturally resistant to feedback inhibition by aromatic amino acids [46,47]. Here, we have selected this gene to be assayed in the engineered strain *S. albidoflavus* UO-FLAV-004, as it lacks the negative feedback inhibition by the final shikimate pathway product (L-tyrosine). In this way, a significant increase in the biosynthesis of diosmetin (1.65-fold) and chrysoeriol (1.34-fold) has been achieved, proving the positive effect on flavonoid biosynthesis of this DAHP synthase gene from *Amycolatopsis balhimycina*, placed in this case under the regulation of a strong constitutive promoter.

This study serves as proof of the suitability of *S. albidoflavus* for the biosynthesis of methylated flavonoids and reveals the efficacy of using a gene coding for a DAHP synthase from a BGC of another actinomycete bacterium, which lacks feedback inhibition by aromatic amino acids (such as the primary flavonoid precursor L-tyrosine), therefore allowing the enhancement of flavonoid titers.

4. Materials and Methods

4.1. Bacterial Strains and Culture Conditions

All strains used in this study are listed in Table 1. *Escherichia coli* TOP10 (Invitrogen, Waltham, MA, USA) was used for routine subcloning. *E. coli* ET12567/pUZ8002 [48] was used for conjugation. The strain used in this study for the heterologous biosynthesis of diosmetin and chrysoeriol was *S. albidoflavus* UO-FLAV-004-NAR [24]. To achieve the heterologous biosynthesis of these two methylated flavones, their biosynthetic pathways were divided into three parts. The first part is the BGC for naringenin biosynthesis, containing the enzymes TAL, 4CL, CHS, and CHI. The genes coding for the enzymes of this BGC were already integrated into the ϕC31 *attb* site of the strain *S. albidoflavus* UO-FLAV-004-NAR. A second plasmid containing the genes coding for the enzymes FNS1 and the chimera F3′H-CPR was integrated into the ϕBT1 *attb* site, giving rise to the strain *S. albidoflavus* UO-FLAV-004-LUT, which is able to produce luteolin. Over this last strain, a third plasmid integration was performed into the pSAM2 site, involving the plasmids pSEVAUO-M31105-M444_29925 or the plasmid pSEVAUO-M31105-At4g26220, generating the strains *S. albidoflavus* UO-FLAV-004-DIO or *S. albidoflavus* UO-FLAV-004-CHR, respectively.

Table 1. Plasmids and strains used in this study.

	Description	Reference
Plasmids		
pSEVA181-At4g26220	Source of *At4g26220* (Level 0 MoClo)	This study
pSEVA181SP25	Source of *SP25* (Level 0 MoClo)	[21]
pSEVA181SP43	Source of *SP43* (Level 0 MoClo)	[21]
pSEVA181-M444_29925	Source of *M444_29925* (Level 0 MoClo)	This study
pSEVA181RiboJ-RBS	Source of *RiboJ-RBS* (Level 0 MoClo)	[21]
pIDTSMARTttsbib	Source of *ttsbib* (Level 0 MoClo)	[21]
pSEVAUO-M21102	Level 2 MoClo receptor	[21]
pSEVAUO-M31205	Level 2 MoClo receptor	[21]
pSEVAUO-M21206F3H-CPR	Level 1 MoClo harboring *F3′H-CPR*	[21]
PCR-Blunt II-TOPO-FNS1	Source of *FNS1* (Level 0 MoClo)	[24]
pSEVAUO-M21102-FNS1	Level 1 MoClo harboring *FNS1*	This study
pSEVAUO-M21503-FNS1/F3′H-CPR	Level 2 MoClo harboring *FNS1* and *F3′H-CPR*	This study
pSEVAUO-M31105-At4g26220	Level 1 MoClo plasmid harboring *At4g26220*	This study
pSEVAUO-M31105-M444_29925	Level 1 MoClo plasmid harboring *M444_29925*	This study
pSEVAUO-M31105	Level 1 MoClo receptor	[21]
pSEVAUO-M31205-dahp	Level 1 MoClo plasmid harboring *dahp*	This study
pSEVAUO-M31505	Level 2 MoClo receptor	[21]
pSEVAUO-M31505-At4g26220-dahp	Level 2 MoClo harboring *At4g26220* and *dahp*	This study
pSEVAUO-M31505-M444_29925-dahp	Level 2 MoClo harboring *M444_29925* and *dahp*	This study
Strains		
E. coli TOP10	Strain used for routine subcloning	Invitrogen (Waltham, MA, USA)
E. coli ET12567/pUZ8002	Strain used for conjugation	[48]
UO-FLAV-004	*S. albidoflavus* strain used in this work	[24]
UO-FLAV-004-NAR	UO-FLAV-004 harboring *TAL, 4CL, CHS* and *CHI*	[24]
UO-FLAV-004-LUT	UO-FLAV-004 harboring *TAL, 4CL, CHS, CHI, FNS1* and *F3′H-CPR*	This study
UO-FLAV-004-DIO	UO-FLAV-004 harboring *TAL, 4CL, CHS, CHI, FNS1, F3′H-CPR* and *M444_29925*	This study
UO-FLAV-004-CHR	UO-FLAV-004 harboring *TAL, 4CL, CHS, CHI, FNS1, F3′H-CPR* and *At4g26220*	This study
UO-FLAV-004-DIO-dahp	UO-FLAV-004 harboring *TAL, 4CL, CHS, CHI, FNS1, F3′H-CPR, M444_29925* and *dahp*	This study
UO-FLAV-004-CHR-dahp	UO-FLAV-004 harboring *TAL, 4CL, CHS, CHI, FNS1, F3′H-CPR, At4g26220* and *dahp*	This study
UO-FLAV-004-FNS1	UO-FLAV-004 harboring FNS1	[24]

Also, two more strains were generated over *S. albidoflavus* UO-FLAV-004-LUT. A MoClo level 2 plasmid was integrated into the pSAM2 chromosomal site containing the genes coding for the M444_29925 *O*-methyltransferase and the DAHP enzyme, generating the strain *S. albidoflavus* UO-FLAV-004-DIO-dahp. In the same manner, a plasmid containing the gene coding for the At4g26220 *O*-methyltransferase plus the DAHP enzyme was integrated in the same position, generating the strain *S. albidoflavus* UO-FLAV-004-CHR-dahp. In order to perform different feeding experiments, the strain *S. albidoflavus* UO-FLAV-004-M444_29925 was generated by transforming the strain *S. albidoflavus* UO-FLAV-004 with the plasmid pSEVAUO-M31105-M444_29925. Finally, the strain *S. albidoflavus* UO-FLAV-004-FNS1 [24] was also used for a feeding experiment.

E. coli strains were grown in tryptic soy broth (TSB, VWR, Barcelona, Spain) or on TSB agar plates, supplemented with the corresponding antibiotics (ampicillin 100 μg/mL, Sigma Aldrich, Madrid, Spain; apramycin 100 μg/mL, Thermo Fisher Scientific, Waltham, MA, USA); kanamycin 100 μg/mL (Alfa Aesar, Karlsruhe, Germany), chloramphenicol 25 μg/mL (AppliChem, Barcelona, Spain), nalidixic acid 50 μg/mL (Acros Organics, Geel, Belgium), and X-gal (AppliChem, Darmstadt, Germany) when blue-white selection was needed. *S. albidoflavus* was grown and sporulated at 30 °C in Bennett medium [49] supplemented with the corresponding antibiotics when necessary (thiostrepton 50 μg/mL, Cayman Chemical, Ann Arbor, MI, USA; hygromycin B 200 μg/mL, Enzo, Barcelona, Spain, or apramycin 50 μg/mL). MA medium (plus 10 mM MgCl2) was used for conjugation between *S. albidoflavus* and *E. coli* [50]. For flavonoid production, *S. albidoflavus* spores were quantified, and an inoculum of 106 spores/mL was performed in triplicate in shake flasks with 25 mL of NL333 medium [51] and incubated for 120 h at 30 °C and 250 rpm.

4.2. Reagents and Biochemicals

All solvents used for solid-phase extraction and HPLC-DAD analysis were LC-MS grade from either Sigma-Aldrich (Madrid, Spain) or VWR Chemicals (Barcelona, Spain). Luteolin, diosmetin, chrysoeriol, eriodictyol, hesperetin, homoeriodictyol, and homohesperetin were provided by Extrasynthese (Genay, France).

4.3. Genes and Enzymes

Restriction enzymes and T4 DNA ligase were purchased from Thermo Fisher Scientific (Madrid, Spain). Synthetic genes for the following ORFs were synthesized by Explora Biotech (Venezia, Italy), after codon optimization: a gene encoding a hypothetical class I SAM-dependent *O*-methyltransferase (M444_29925) from *Streptomyces* sp. Mg1 (Genbank accession no. OR820610); a gene encoding a 3-deoxy-7-phosphoheptulonate synthase (dahp) from *Amycolatopsis balhimycina* (Genbank accession no. OR820611); *At4g26220* from *Arabidopsis thaliana* (Genbank accession no. OR820609). Other genes used in this study were WP_013066811 (for *TAL*), NP_628552 (for *4CL*), L07647.1 (*CHS*), AY595413.1 (*CHI*), AY230247.1 (*FNS*), OQ674225 (*3′FH/CPR*).

4.4. Plasmids Construction

All the plasmids used in this study are listed in Table 1. Final constructs are depicted in Figure S5.

4.4.1. Construction of pSEVAUO-M21503-FNS1/F3′H-CPR

The gene encoding FNS1 was assembled in a level 1 MoClo reaction using the level 0 plasmids PCR-Blunt II-TOPO-FNS1 [24] pSEVA181SP43, pSEVA181RiboJ-RBS, pIDTSMARTttsbib [21], and the level 1 MoClo receptor pSEVAUO-M21102 [21], yielding the plasmid pSEVAUO- M21102–FNS1. Then, this plasmid was used in combination with the plasmid pSEVAUO-M21206F3H-CPR and the level 2 MoClo receptor pSEVAUO-M21503 [21] to generate the plasmid pSEVAUO-M21503-FNS1/F3′H-CPR.

4.4.2. Construction of pSEVAUO-M31105-At4g26220, pSEVAUO-M31105-M444_29925, and pSEVAUO-M31205-dahp

pSEVAUO-M31105-At4g26220 and pSEVAUO-M31105-M444_29925 are level 1 MoClo plasmids, and they were assembled by combining the level 1 MoClo receptor pSEVAUO-M31105 and the level 0 plasmids pSEVA181SP25, pSEVA181RiboJ-RBS, pIDTSMARTtts-bib [21], and pSEVA181-At4g26220 (this study) or pSEVA181-M444_29925 (this study), respectively.

The plasmid pSEVAUO-M31205-dahp was assembled using the level 1 MoClo receptor pSEVAUO-M31205 and the level 0 plasmids pSEVA181SP25, pSEVA181RiboJ-RBS, pIDTSMARTttsbib, and pSEVA181-dahp (this study).

4.4.3. Construction of pSEVAUO-M31505-At4g26220-dahp and pSEVAUO-M31505-M444_29925-dahp

The *At4g26220* and *M444_29925* genes were brought together with the *dahp* gene in two level 2 MoClo reactions. The first reaction was performed with the plasmids pSEVAUO-M31105-At4g26220, pSEVAUO-M31205-dahp, and the level 2 MoClo receptor pSEVAUO-M31505 [21], resulting in pSEVAUO-M31505-At4g26220-dahp. The second reaction was performed using the plasmids pSEVAUO-M31105-M444_29925 and pSEVAUO-M31205-dahp, yielding the plasmid pSEVAUO-M31505-M444_29925-dahp.

4.5. Flavonoid Extraction and LC-DAD Analysis

Spores from the different *S. albidoflavus* strains were incubated, as described before. Flavonoids were recuperated by organic extraction with acetone (cellular pellet) and ethyl acetate (culture supernatant). A sample of 1 mL was taken from the flasks and centrifuged at 12,000 rpm for 1 min to separate the supernatant from the pellet. The pellet was extracted with 1 mL of acetone using vortex for 1 h. The supernatant was extracted with 800 μL of ethyl acetate by agitation for 10 min. Both pellet and supernatant extractions were centrifuged for 1 min at 12,000 rpm, and the organic fractions were mixed and dried in a speed-vac. A second extraction was performed using 800 μL ethyl acetate over the cellular pellet and the supernatant using vortex and agitation, respectively, as described before. Finally, these extractions (cellular pellet and supernatant) were mixed with the initial dry extract obtained in the first extraction and dried in a speed-vac.

For the identification of flavonoids using HPLC-DAD, the final dry extract obtained from each cultivation condition was dissolved in 100 μL DMSO/MeOH 1:1 (*v*/*v*), and the samples were centrifuged prior to injection into the equipment. The HPLC separation was performed on an HPLC (1260 Infinity, Agilent Technologies, Madrid, Spain) equipped with an analytical column Pursuit XRs C18 (50 × 4.0 mm, 5 μm, Agilent Technologies, Madrid, Spain). HPLC gradient was made with analytical grade solvent B (acetonitrile (VWR, Barcelona, Spain), and water as solvents (1 mL/min flow rate). All solvents contained 0.1% formic acid. Samples were run by an isocratic elution of 10% MeCN from 0 min to 5.44 min, followed by a linear gradient from 10% to 35% of MeCN from min 5.44 to min 21.77, maintaining the mobile phase composition until 27.21 min. Then, a linear gradient from 35% to 100% MeCN between 27.21 min and 43.54 min was applied, followed by an isocratic elution until 55 min. Then, a linear gradient from 100% to 10% MeCN was applied from 55 min to 56 min. Finally, this mobile phase composition was maintained until the end of the program (61). Detection and spectral characterization of peaks were carried out with a photodiode array detector, and the analysis was performed with Data Analysis 4.3 software (Bruker, Billerica, MA, USA). All chromatograms were extracted at 280 nm. The column temperature was set to 30 °C. Flavonoids luteolin, diosmetin, chrysoeriol, hesperetin, homoeriodictyol, and homohesperetin were identified using authentic commercial standards. Luteolin, diosmetin, chrysoeriol, and hesperetin were quantified by comparing the peak area with that of a known amount of an authentic compound through a calibration curve. The production titers are expressed in mg/L, and the mean value was calculated from three biological replicates.

4.6. Statistical Analysis

Two-way ANOVA (analysis of variance) with Sidak's multiple comparisons test was used to test the differences in the biosynthesis of diosmetin among the strains *S. albidoflavus* UO-FLAV-004-DIO and *S. albidoflavus* UO-FLAV-004-DIO-dahp and the biosynthesis of chrysoeriol among the strain *S. albidoflavus* UO-FLAV-004-CHR and the strain *S. albidoflavus* UO-FLAV-004-CHR-dahp. Graphical representation of the different generated data was carried out using GraphPad Prism software (version 9.0.2, GraphPad Software, San Diego, CA, USA), with a p-value < 0.05 considered as statistically significant (* $p < 0.05$; ** $p < 0.005$; *** $p < 0.0005$; **** $p < 0.0001$).

Supplementary Materials: The following supporting information can be downloaded at: https://www.mdpi.com/article/10.3390/ijms25052776/s1.

Author Contributions: Funding acquisition (F.L.), investigation (Á.P.-V. and J.S.-D.); supervision (C.J.V. and F.L.); writing—original draft (Á.P.-V.); writing—review and editing (Á.P.-V., C.J.V., J.S.-D. and F.L.). All authors have read and agreed to the published version of the manuscript.

Funding: This research was funded by Principado de Asturias (Spain) through the program "Ayudas a organismos públicos para apoyar las actividades de I + D + I de sus grupos de investigación" (grant AYUD/2021/51347), as well as by the "Programa Severo Ochoa de Ayudas Predoctorales para la investigación y docencia" from Principado de Asturias (PhD grant PA-21-PF-BP20-150 to Á.P.-V.), "Programa de Ayudas FPI" from MICINN (PhD grant PRE2022-102792 to J.S.-D.), the research project PID2021-127812OB-I00 from MICINN (Spanish Ministry of Science and Innovation), and the European Union's Horizon 2020 Research and Innovation Program under Grant Agreement no. 814650 for the project SynBio4Flav.

Institutional Review Board Statement: Not applicable.

Informed Consent Statement: All authors have read and approved the final version of the manuscript and have accepted its publication in this journal.

Data Availability Statement: Data and materials can be obtained from the research group upon request. Sequences accession numbers have been included in the Section 4.

Acknowledgments: The authors thank Extrasynthese (Genay, France) for providing flavonoid standards for HPLC-MS analyses. The authors thank Explora Biotech (Venezia, Italy) for providing synthetic genes.

Conflicts of Interest: The authors declare no conflicts of interest.

References

1. Kumar, S.; Pandey, A.K. Chemistry and Biological Activities of Flavonoids: An Overview. *Sci. World J.* **2013**, *2013*, 162750. [CrossRef] [PubMed]
2. Li, A.-N.; Li, S.; Zhang, Y.-J.; Xu, X.-R.; Chen, Y.-M.; Li, H.-B. Resources and Biological Activities of Natural Polyphenols. *Nutrients* **2014**, *6*, 6020–6047. [CrossRef] [PubMed]
3. Shah, F.L.A.; Ramzi, A.B.; Baharum, S.N.; Noor, N.M.; Goh, H.H.; Leow, T.C.; Oslan, S.N.; Sabri, S. Recent Advancement of Engineering Microbial Hosts for the Biotechnological Production of Flavonoids. *Mol. Biol. Rep.* **2019**, *46*, 6647–6659. [CrossRef]
4. Al-Khayri, J.M.; Sahana, G.R.; Nagella, P.; Joseph, B.V.; Alessa, F.M.; Al-Mssallem, M.Q. Flavonoids as Potential Anti-Inflammatory Molecules: A Review. *Molecules* **2022**, *27*, 2901. [CrossRef]
5. Kaushal, N.; Singh, M.; Singh Sangwan, R. Flavonoids: Food Associations, Therapeutic Mechanisms, Metabolism and Nanoformulations. *Food Res. Int.* **2022**, *157*, 111442. [CrossRef]
6. Roy, A.; Khan, A.; Ahmad, I.; Alghamdi, S.; Rajab, B.S.; Babalghith, A.O.; Alshahrani, M.Y.; Islam, S.; Islam, M.R. Flavonoids a Bioactive Compound from Medicinal Plants and Its Therapeutic Applications. *BioMed Res. Int.* **2022**, *2022*, 5445291. [CrossRef]
7. Zhao, L.; Yuan, X.; Wang, J.; Feng, Y.; Ji, F.; Li, Z.; Bian, J. A Review on Flavones Targeting Serine/Threonine Protein Kinases for Potential Anticancer Drugs. *Bioorg. Med. Chem.* **2019**, *27*, 677–685. [CrossRef]
8. Zhao, K.; Yuan, Y.; Lin, B.; Miao, Z.; Li, Z.; Guo, Q.; Lu, N. LW-215, a Newly Synthesized Flavonoid, Exhibits Potent Anti-Angiogenic Activity In Vitro and In Vivo. *Gene* **2018**, *642*, 533–541. [CrossRef]
9. Camero, C.M.; Germanò, M.P.; Rapisarda, A.; D'Angelo, V.; Amira, S.; Benchikh, F.; Braca, A.; De Leo, M. Anti-Angiogenic Activity of Iridoids from *Galium tunetanum*. *Rev. Bras. Farmacogn.* **2018**, *28*, 374–377. [CrossRef]

10. Patel, K.; Kumar, V.; Rahman, M.; Verma, A.; Patel, D.K. New Insights into the Medicinal Importance, Physiological Functions and Bioanalytical Aspects of an Important Bioactive Compound of Foods 'Hyperin': Health Benefits of the Past, the Present, the Future. *Beni-Suef Univ. J. Basic Appl. Sci.* **2018**, *7*, 31–42. [CrossRef]
11. Wen, X.; Walle, T. Methylated Flavonoids Have Greatly Improved Intestinal Absorption and Metabolic Stability. *Drug Metab. Dispos.* **2006**, *34*, 1786–1792. [CrossRef]
12. Patel, K.; Gadewar, M.; Tahilyani, V.; Patel, D.K. A Review on Pharmacological and Analytical Aspects of Diosmetin: A Concise Report. *Chin. J. Integr. Med.* **2013**, *19*, 792–800. [CrossRef]
13. Chan, B.C.L.; Ip, M.; Gong, H.; Lui, S.L.; See, R.H.; Jolivalt, C.; Fung, K.P.; Leung, P.C.; Reiner, N.E.; Lau, C.B.S. Synergistic Effects of Diosmetin with Erythromycin against ABC Transporter Over-Expressed Methicillin-Resistant *Staphylococcus aureus* (MRSA) RN4220/PUL5054 and Inhibition of MRSA Pyruvate Kinase. *Phytomedicine* **2013**, *20*, 611–614. [CrossRef]
14. Aboulaghras, S.; Sahib, N.; Bakrim, S.; Benali, T.; Charfi, S.; Guaouguaou, F.E.; El Omari, N.; Gallo, M.; Montesano, D.; Zengin, G.; et al. Health Benefits and Pharmacological Aspects of Chrysoeriol. *Pharmaceuticals* **2022**, *15*, 973. [CrossRef] [PubMed]
15. Shen, N.; Wang, T.; Gan, Q.; Liu, S.; Wang, L.; Jin, B. Plant Flavonoids: Classification, Distribution, Biosynthesis, and Antioxidant Activity. *Food Chem.* **2022**, *383*, 132531. [CrossRef] [PubMed]
16. Tsao, R. Chemistry and Biochemistry of Dietary Polyphenols. *Nutrients* **2010**, *2*, 1231–1246. [CrossRef]
17. Falcone Ferreyra, M.L.; Rius, S.P.; Casati, P. Flavonoids: Biosynthesis, Biological Functions, and Biotechnological Applications. *Front. Plant Sci.* **2012**, *3*, 222. [CrossRef] [PubMed]
18. Chouhan, S.; Sharma, K.; Zha, J.; Guleria, S.; Koffas, M.A.G. Recent Advances in the Recombinant Biosynthesis of Polyphenols. *Front. Microbiol.* **2017**, *8*, 2259. [CrossRef] [PubMed]
19. Wang, Y.; Chen, S.; Yu, O. Metabolic Engineering of Flavonoids in Plants and Microorganisms. *Appl. Microbiol. Biotechnol.* **2011**, *91*, 949–956. [CrossRef] [PubMed]
20. Falcone Ferreyra, M.L.; Emiliani, J.; Rodriguez, E.J.; Campos-Bermudez, V.A.; Grotewold, E.; Casati, P. The Identification of Maize and Arabidopsis Type I FLAVONE SYNTHASEs Links Flavones with Hormones and Biotic Interactions. *Plant Physiol.* **2015**, *169*, 1090–1107. [CrossRef] [PubMed]
21. Magadán-Corpas, P.; Ye, S.; Pérez-Valero, Á.; McAlpine, P.L.; Valdés-Chiara, P.; Torres-Bacete, J.; Nogales, J.; Villar, C.J.; Lombó, F. Optimized De Novo Eriodictyol Biosynthesis in *Streptomyces albidoflavus* Using an Expansion of the Golden Standard Toolkit for Its Use in Actinomycetes. *Int. J. Mol. Sci.* **2023**, *24*, 8879. [CrossRef]
22. Kim, M.J.; Kim, B.G.; Ahn, J.H. Biosynthesis of Bioactive O-Methylated Flavonoids in *Escherichia coli*. *Appl. Microbiol. Biotechnol.* **2013**, *97*, 7195–7204. [CrossRef] [PubMed]
23. Liu, X.; Cheng, J.; Zhu, X.; Zhang, G.; Yang, S.; Guo, X.; Jiang, H.; Ma, Y. De Novo Biosynthesis of Multiple Pinocembrin Derivatives in *Saccharomyces cerevisiae*. *ACS Synth. Biol.* **2020**, *9*, 3042–3051. [CrossRef] [PubMed]
24. Pérez-Valero, Á.; Ye, S.; Magadán-Corpas, P.; Villar, C.J.; Lombó, F. Metabolic Engineering in *Streptomyces albidoflavus* for the Biosynthesis of the Methylated Flavonoids Sakuranetin, Acacetin, and Genkwanin. *Microb. Cell Factories* **2023**, *22*, 234. [CrossRef] [PubMed]
25. Sharma, V.; Kaur, R.; Salwan, R. *Streptomyces: Host for Refactoring of Diverse Bioactive Secondary Metabolites*; Springer International Publishing: Cham, Switzerland, 2021; Volume 11, ISBN 0123456789.
26. Kuhstoss, S.; Rao, R.N. Analysis of the Integration Function of the Streptomycete Bacteriophage ΦC31. *J. Mol. Biol.* **1991**, *222*, 897–908. [CrossRef] [PubMed]
27. Gregory, M.A.; Till, R.; Smith, M.C.M. Integration Site for *Streptomyces* Phage ΦBT1 and Development of Site-Specific Integrating Vectors. *J. Bacteriol.* **2003**, *185*, 5320–5323. [CrossRef] [PubMed]
28. Raynal, A.; Friedmann, A.; Tuphile, K.; Guerineau, M.; Pernodet, J.L. Characterization of the AttP Site of the Integrative Element PSAM2 from *Streptomyces ambofaciens*. *Microbiology* **2002**, *148*, 61–67. [CrossRef]
29. Thykaer, J.; Nielsen, J.; Wohlleben, W.; Weber, T.; Gutknecht, M.; Lantz, A.E.; Stegmann, E. Increased Glycopeptide Production after Overexpression of Shikimate Pathway Genes Being Part of the Balhimycin Biosynthetic Gene Cluster. *Metab. Eng.* **2010**, *12*, 455–461. [CrossRef]
30. Ikeda, M. Towards Bacterial Strains Overproducing L-Tryptophan and Other Aromatics by Metabolic Engineering. *Appl. Microbiol. Biotechnol.* **2006**, *69*, 615–626. [CrossRef]
31. Pandurangan, N. A New Synthesis for Acacetin, Chrysoeriol, Diosmetin, Tricin and Other Hydroxylated Flavones by Modified Baker-Venkataraman Transformation. *Lett. Org. Chem.* **2014**, *11*, 225–229. [CrossRef]
32. Victor, M.M.; David, J.M.; Cortez, M.V.M.; Leite, J.L.; da Silva, G.S.B. A High-Yield Process for Extraction of Hesperidin from Orange (*Citrus sinensis* L. Osbeck) Peels Waste, and Its Transformation to Diosmetin, A Valuable and Bioactive Flavonoid. *Waste Biomass Valorization* **2021**, *12*, 313–320. [CrossRef]
33. Myronovskyi, M.; Luzhetskyy, A. Native and Engineered Promoters in Natural Product Discovery. *Nat. Prod. Rep.* **2016**, *33*, 1006–1019. [CrossRef] [PubMed]
34. Darsandhari, S.; Dhakal, D.; Shrestha, B.; Parajuli, P.; Seo, J.-H.; Kim, T.-S.; Sohng, J.K. Characterization of Regioselective Flavonoid O-Methyltransferase from the *Streptomyces* Sp. KCTC 0041BP. *Enzym. Microb. Technol.* **2018**, *113*, 29–36. [CrossRef] [PubMed]
35. Wils, C.R.; Brandt, W.; Manke, K.; Vogt, T. A Single Amino Acid Determines Position Specificity of an *Arabidopsis thaliana* CCoAOMT-like O-Methyltransferase. *FEBS Lett.* **2013**, *587*, 683–689. [CrossRef]

36. Schröder, G.; Wehinger, E.; Lukačin, R.; Wellmann, F.; Seefelder, W.; Schwab, W.; Schröder, J. Flavonoid Methylation: A Novel 4′-O-Methyltransferase from Catharanthus Roseus, and Evidence That Partially Methylated Flavanones Are Substrates of Four Different Flavonoid Dioxygenases. *Phytochemistry* **2004**, *65*, 1085–1094. [CrossRef]
37. Aisa, H.A.; Izotova, L.; Karimov, A.; Botirov, E.; Mamadrahimov, A.; Ibragimov, B. Crystal, Molecular Structure and Hirshheld Surface Analysis of 5-Hydroxy-3,6,7,8-Tetramethoxyflavone. *Acta Crystallogr. Sect. E Crystallogr. Commun.* **2020**, *76*, 1748–1751. [CrossRef]
38. Shin, W.; Lah, M.S. Structure of (R,S)-Naringenin. *Acta Crystallogr. Sect. C Cryst. Struct. Commun.* **1986**, *42*, 626–628. [CrossRef]
39. Lütke-Eversloh, T.; Santos, C.N.S.; Stephanopoulos, G. Perspectives of Biotechnological Production of L-Tyrosine and Its Applications. *Appl. Microbiol. Biotechnol.* **2007**, *77*, 751–762. [CrossRef]
40. Koopman, F.; Beekwilder, J.; Crimi, B.; Van Houwelingen, A.; Hall, R.D.; Bosch, D.; Van Maris, A.J.A.; Pronk, J.T.; Daran, J. De Novo Production of the Flavonoid Naringenin in Engineered *Saccharomyces cerevisiae*. *Microb. Cell Factories* **2012**, *11*, 155. [CrossRef]
41. Wu, J.; Zhou, T.; Du, G.; Zhou, J.; Chen, J. Modular Optimization of Heterologous Pathways for de Novo Synthesis of (2S)-Naringenin in *Escherichia coli*. *PLoS ONE* **2014**, *9*, e101492. [CrossRef]
42. Zhou, S.; Hao, T.; Zhou, J. Fermentation and Metabolic Pathway Optimization to De Novo Synthesize (2S)-Naringenin in *Escherichia coli*. *J. Microbiol. Biotechnol.* **2020**, *30*, 1574–1582. [CrossRef]
43. Lee, H.; Kim, B.G.; Kim, M.; Ahn, J.H. Biosynthesis of Two Flavones, Apigenin and Genkwanin, in *Escherichia coli*. *J. Microbiol. Biotechnol.* **2015**, *25*, 1442–1448. [CrossRef] [PubMed]
44. August, P.R.; Tang, L.; Yoon, Y.J.; Ning, S.; Müller, R.; Yu, T.-W.; Taylor, M.; Hoffmann, D.; Kim, C.-G.; Zhang, X.; et al. Biosynthesis of the Ansamycin Antibiotic Rifamycin: Deductions from the Molecular Analysis of the Rif Biosynthetic Gene Cluster of Amycolatopsis Mediterranei S699. *Chem. Biol.* **1998**, *5*, 69–79. [CrossRef] [PubMed]
45. He, J.; Magarvey, N.; Piraee, M.; Vining, L.C. The Gene Cluster for Chloramphenicol Biosynthesis in *Streptomyces venezuelae* ISP5230 Includes Novel Shikimate Pathway Homologues and a Monomodular Non-Ribosomal Peptide Synthetase Gene The GenBank Accession Number for the Sequence Reported in This Paper is AF262220. *Microbiology* **2001**, *147*, 2817–2829. [CrossRef]
46. Dyer, W.E.; Weaver, L.M.; Zhao, J.M.; Kuhn, D.N.; Weller, S.C.; Herrmann, K.M. A CDNA Encoding 3-Deoxy-D-Arabino-Heptulosonate 7-Phosphate Synthase from *Solanum tuberosum* L. *J. Biol. Chem.* **1990**, *265*, 1608–1614. [CrossRef] [PubMed]
47. Pinto, J.E.B.P.; Suzich, J.A.; Herrmann, K.M. 3-Deoxy-d- Arabino -Heptulosonate 7-Phosphate Synthase from Potato Tuber (*Solanum tuberosum* L.). *Plant Physiol.* **1986**, *82*, 1040–1044. [CrossRef] [PubMed]
48. Macneil, D.J.; Gewain, K.M.; Ruby, C.L.; Dezeny, G.; Gibbons, P.H.; Maeneil, T. Analysis of *Streptomyces avermitilis* Genes Required for Avermectin Biosynthesis Utilizing a Novel Inte-Gration Vector. *Gene* **1992**, *111*, 61–68. [CrossRef] [PubMed]
49. Kieser, T.; Bibb, M.J.; Buttner, M.J.; Chater, K.F.; Hopwood, D.A.; John Innes Foundation. *Practical Streptomyces Genetics*; John Innes Foundation: Norwich, UK, 2000; Volume 291.
50. Fernández, E.; Weissbach, U.; Sánchez Reillo, C.; Braña, A.F.; Méndez, C.; Rohr, J.; Salas, J.A. Identification of Two Genes from *Streptomyces argillaceus* Encoding Glycosyltransferases Involved in Transfer of a Disaccharide during Biosynthesis of the Antitumor Drug Mithramycin. *J. Bacteriol.* **1998**, *180*, 4929–4937. [CrossRef]
51. Myronovskyi, M.; Tokovenko, B.; Brötz, E.; Rückert, C.; Kalinowski, J.; Luzhetskyy, A. Genome Rearrangements of Streptomyces Albus J1074 Lead to the Carotenoid Gene Cluster Activation. *Appl. Microbiol. Biotechnol.* **2014**, *98*, 795–806. [CrossRef]

International Journal of
Molecular Sciences

MDPI

Article

Efficient Production of 9,22-Dihydroxy-23,24-bisnorchol-4-ene-3-one from Phytosterols by Modifying Multiple Genes in *Mycobacterium fortuitum*

Suwan Han [1,2], Xiangcen Liu [1,2], Beiru He [1,2,3], Xinghui Zhai [1,2], Chenyang Yuan [1,2,3], Yixin Li [4], Weichao Lin [1], Haoyu Wang [3] and Baoguo Zhang [1,2,*]

1 Laboratory of Biorefinery, Shanghai Advanced Research Institute, Chinese Academy of Sciences, No. 99 Haike Road, Pudong, Shanghai 201210, China; hansw@sari.ac.cn (S.H.); liucx@sari.ac.cn (X.L.); hebr2022@shanghaitech.edu.cn (B.H.); zhaixh@sari.ac.cn (X.Z.); yuanchy@shanghaitech.edu.cn (C.Y.); linwc@sari.ac.cn (W.L.)
2 University of Chinese Academy of Sciences, Beijing 100049, China
3 School of Life Science and Technology, ShanghaiTech University, Shanghai 201210, China; wanghy2022@shanghaitech.edu.cn
4 Department of Biology, Colby College, Waterville, ME 04901, USA; yli24@colby.edu
* Correspondence: zhangbg@sari.ac.cn

Abstract: C19 steroids and C22 steroids are vital intermediates for the synthesis of steroid drugs. Compared with C19 steroids, C22 steroids are more suitable for synthesizing progesterone and adrenocortical hormones, albeit less developed. 9,22-dihydroxy-23,24-bisnorchol-4-ene-3-one(9-OHBA), due to its substituents at positions C-9 and C-22, is a beneficial and innovative steroid derivative for synthesizing corticosteroids. We focused on the C22 pathway in *Mycobacterium fortuitum* ATCC 35855, aiming to develop a productive strain that produces 9-OHBA. We used a mutant strain, MFΔ*kstD*, that knocked out *kstds* from *Mycobacterium fortuitum* ATCC 35855 named MFKD in this study as the original strain. Hsd4A and FadA5 are key enzymes in controlling the C19 metabolic pathway of steroids in *Mycobacterium fortuitum* ATCC 35855. After knocking out *hsd4A*, MFKDΔ*hsd4A* accumulated 81.47% 9-OHBA compared with 4.13% 9-OHBA in the strain MFKD. The double mutant MFKDΔ*hsd4A*Δ*fadA5* further improved the selectivity of 9-OHBA to 95.13%, and 9α-hydroxy-4-androstenedione (9-OHAD) decreased to 0.90% from 4.19%. In the end, we obtained 6.81 g/L 9-OHBA from 10 g/L phytosterols with a molar yield of 80.33%, which showed the best performance compared with formerly reported strains.

Keywords: C22 steroids; 9,22-dihydroxy-23,24-bisnorchol-4-enehp-3-one(9-OHBA); *Mycobacterium fortuitum*; phytosterol; *hsd4A*; *fadA5*

1. Introduction

Steroid-based drugs have a wide range of therapeutic uses, including regulating hormone levels and blood pressure and suppressing the inflammatory response [1–3]. C19 and C22 intermediates can be converted into steroid-based drugs using chemical methods [4–6]. Nowadays, the biotransformation of steroid-based drug intermediates from phytosterols attracts more attention due to its eco-friendliness and mild reaction conditions compared to the traditional chemical synthesis routine [7].

Phytosterols, consisting of β-sitosterol, campesterol, and stigmasterol, are accessible on the raw materials market and are often produced as a by-product or waste of the vegetable oil refining industry [8]. Thanks to their low cost and abundant availability, phytosterols are good substitutes for producing steroid-based drug intermediates. However, β-sitosterol, stigmasterol, and campesterol are only distinguished by one carbon atom or one double bond on the side chain, which brings about high costs for the separation of each component. Fortunately,

Actinomycetes, especially mycobacteria, known for their outstanding abilities of incompletely degrading the different steroid components in phytosterols, have been industrially used to produce steroid-based drug intermediates and have also been genetically modified to accumulate steroid-based drug intermediates. Among these, technologies for the production of C19 intermediates consisting of 4-androstene-3,17-dione (AD), 1,4-androstadiene-3,17-dione (ADD), and 9α-hydroxy-4-androstenedione (9-OHAD) have been studied a lot [9–14]. By contrast, technologies for the production of C22 intermediates are less developed, which show advantages in synthesizing some progestin and adrenal hormones, such as progesterone, hydrocortisone, dexamethasone, and drospirenone [15–17].

The C22 steroid intermediates include 22-hydroxy-23,24-bisnorchol-4-en-3-one (BA), 22-hydroxy-23,24-bisnorchol-1,4-dien-3-one (1,4-BA), and 9,22-dihydroxy-23,24-bisnorchol-4-en-3-one (9-OHBA). Among these, 9-OHBA is a beneficial and innovative steroid derivative for synthesizing corticosteroids due to its hydroxyl group at positions C-9 and C-22. Using *M. neoaurum* ATCC 25795 (NwIB-XII) as a model, the deletion of *hsd4A* resulted in the production of 23,24-bisnorcholenic steroids [17]. Finally, *M. neoaurum* ATCC 25795 deleting *hsd4A*, *kstD1*, *kstD2*, and *kstD3* accumulated 10.25–10.72 g 9-OHBA from 40 g phytosterols with a molar yield of 30–32% [17]. Later in 2021, the double deletion of *hsd4A* and *fadA5* in a *M. neoaurum* DSM 44074 *kstd*-null strain was found to significantly accumulate 3.58 g/L 9-OHBA with the cultivation of 5 g/L phytosterols [18]. However, when the phytosterol concentration rose to 10 g/L, the concentration of 9-OHBA sharply decreased to 2.73 g/L. Therefore, a strain with a higher yield and selectivity of 9-OHBA needs to be developed.

In our previous work, we solved the problem of the degradation of 9-OHBA and 9-OHAD by knocking all putative *kstds* in *Mycobacterium fortuitum* ATCC 35855 named MFKD in this study [19]. The mutant strain MFKD accumulated mainly 9-OHAD and 9-OHBA as a by-product. Therefore, to construct a 9-OHBA-producing strain, we tried to identify the key enzymes, as shown in Figure 1, that control metabolic flow between the C22 and C19 pathways. We bioinformatically identified one *hsd4A* gene, one *fadA5* gene, and one *opccR* gene in *Mycobacterium fortuitum* ATCC 35855 that were mainly responsible for switching the C19 and C22 pathways. After multiple genetic modifications, the results showed that blocking the C19 pathway increased the production of 9-OHBA. Our results also contributed to the knowledge of the complexity and diversity associated with regulating steroid catabolism in *M. fortuitum*. They provided a theoretical basis for the optimization of industrial microbial biocatalysts.

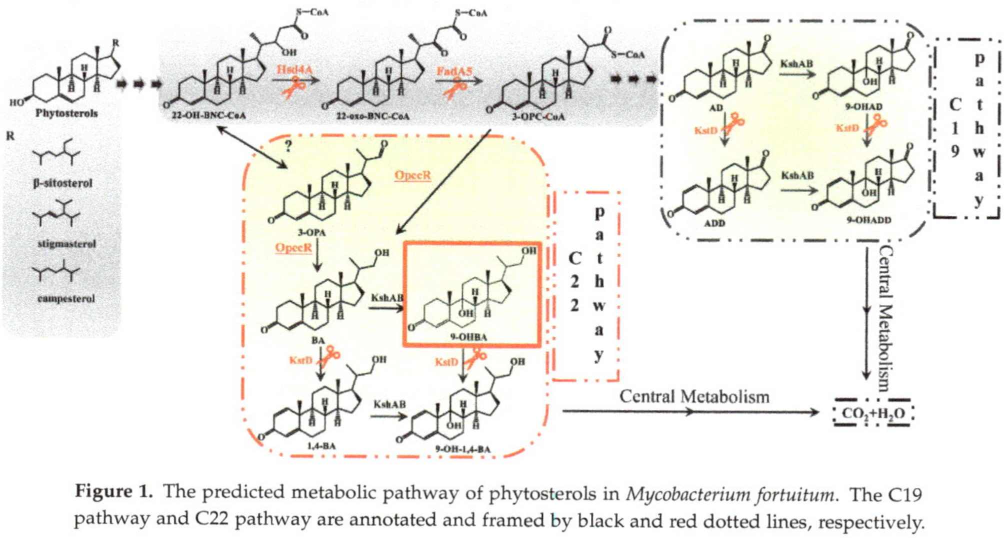

Figure 1. The predicted metabolic pathway of phytosterols in *Mycobacterium fortuitum*. The C19 pathway and C22 pathway are annotated and framed by black and red dotted lines, respectively.

All edited genes are emphasized in red color. The question mark means that the enzyme catalyzing this reaction is not yet known in *Mycobacterium fortuitum*. Phytosterols consist of β-sitosterol, campesterol, and stigmasterol, which differ in their R groups. The diagram was drawn based on our previous article [16]. 22-OH-BNC-CoA, 22-hydroxy-3,24-dioxo-4-ene-cholest-CoA; 22-oxo-BNC-CoA, 3,22,24-trioxo-4-ene-cholest-CoA; 3-OPC-CoA, 3,22-dioxo-4-ene-pregna-CoA; 3-OPA, 3-oxo-4-ene-pregna-20-carboxyaldehyde; BA, 22-hydroxy-23, 24-bisnorchol-4-en-3-one; 1,4-BA, 22-hydroxy-23,24-bisnorchol-1,4-dien-3-one; 9-OHBA, 9,22-dihydroxy-23,24-bisnorchol-4-ene-3-one; AD, 4-androstene-3,17-dione; ADD, 1,4-androstadiene-3,17-dione; 9-OHAD, 9α-hydroxy-4-androstenedione; KshAB, 3-ketosteroid-9α-hydroxylase; KstD, 3-ketosteroid-Δ^1-dehydrogenase; OpccR, a dual-function reductase; Hsd4A, 17β-hydroxysteroid dehydrogenase/β-hydroxyacyl CoA dehydrogenase; FadA5, acetyl-CoA-acetyltransferase/thiolase.

2. Results

2.1. Construction of Phylogenetic Tree of Hsd4A, FadA5, and OpccR

Whole-genome sequencing identified one putative *hsd4A*(gene1030), one putative *fadA5*(gene5404), and one putative *opccR* (gene1719) in *M. fortuitum* ATCC 35855. We translated the nucleotide sequences of the genes into amino acid sequences, and then, aligned them with their homologues, whose physicochemical properties and role in steroid metabolism were examined. The amino acid sequence of Hsd4A in *M. fortuitum* ATCC 35855 shared an identity of 78.8% with Hsd4A in *M. neoaurum* DSM 44074 [18], 78.5% with Hsd4A in *M. neoaurum* HGMS2 [20], 77.5% with Hsd4A in *M. neoaurum* DSM 1381 [21], and 69.3% with Hsd4A in *M. tuberculosis* H37V [13], whose Hsd4A enzymes were reported to have catalytic activities. As shown in Figure 2, their amino acid sequences shared a high similarity, while Hsd4A in *M. fortuitum* ATCC 35855 did not, which suggested a new unidentified Hsd4A.

The amino acid sequence of FadA5 of *M. fortuitum* ATCC 35855 shared a high identity of 88.9%, 88.4%, and 81.6% with FadA5 of *M. neoaurum* HGMS2 [20], *M. neoaurum* DSM 44074 [18], and *M. tuberculosis* H37V [12,22], respectively. Moreover, the alignments of the amino acid sequence of FadA5 confirmed that FadA5 has a conserved set of catalytic residues. The key amino acids at the active site, including Cys93, acted as nucleophiles attacking the steroid acyl-CoA's β-keto carbonyl moiety. Additionally, Cys377 and His347 serve as regular acid/base residues. These results were consistent with Lu's research in 2017 [22].

According to previous reports, OpccR was found in both *M. neoaurum* HGMS2 and *M. neoaurum* CCTCC AB2019054 and was reported to have catalytic activity in the steroid metabolic pathway [5,20]. One putative OpccR in *M. fortuitum* ATCC 35855 was found by sequence alignment, and shared 76.3% identity with those in *M. neoaurum* HGMS2 and *M. neoaurum* CCTCC AB2019054. A dendrogram of OpccR was also built and was chosen to analyze and speculate the function of Opccr in steroid metabolism. The alignment results revealed that the NADPH-binding motif in the C terminal domain (S375AA-G377AA-R398AA) was conserved.

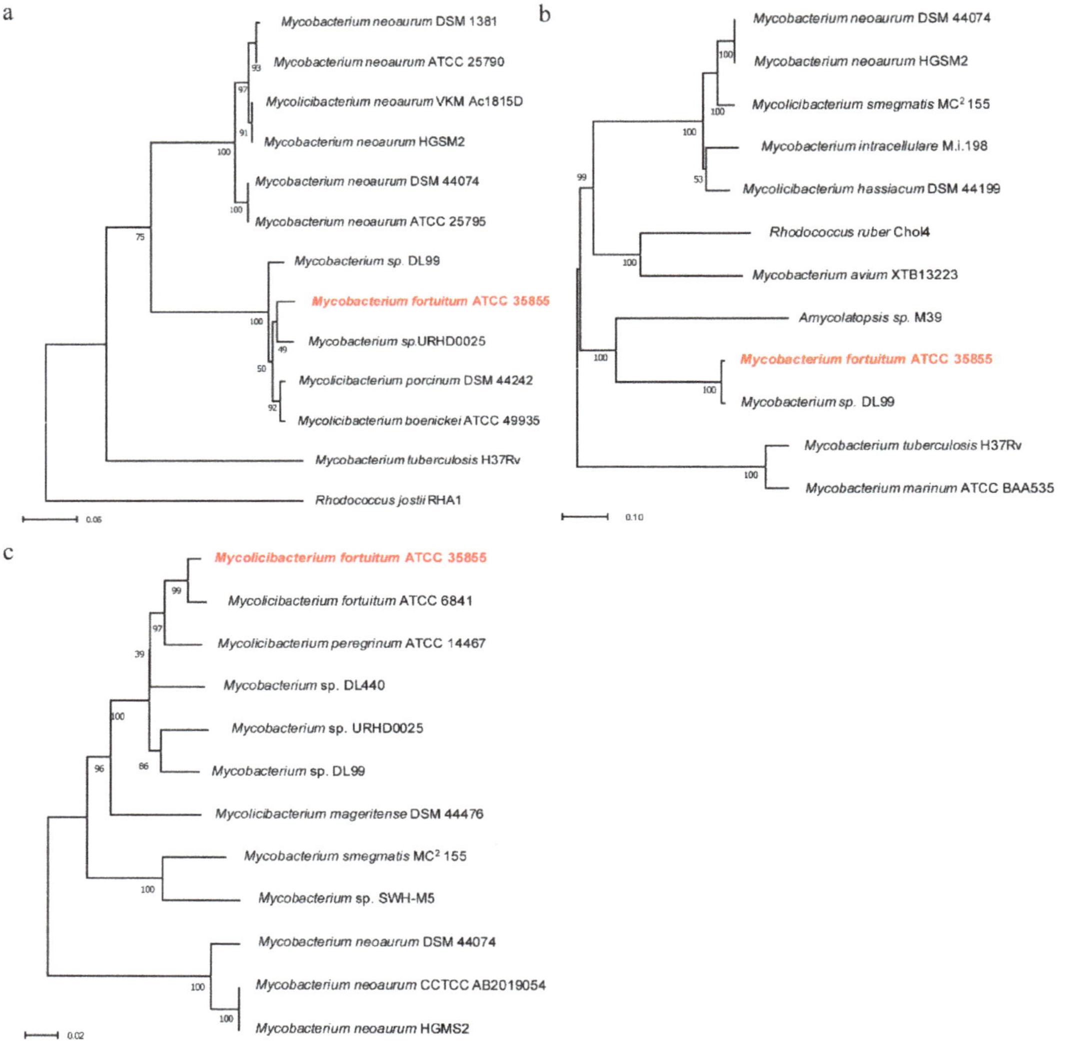

Figure 2. Phylogenetic trees of Hsd4A, FadA5, and OpccR in *Mycobacterium fortuitum* ATCC 35855 and other representative homologues. *Mycobacterium fortuitum* ATCC 35855 is highlighted in red color. (**a**) Phylogenetic tree of Hsd4A; (**b**) phylogenetic tree of FadA5; (**c**) phylogenetic tree of OpccR.

2.2. Hsd4A—The Key Enzyme in the C19 Pathway

As is shown in Figure 1, *hsd4A* controls the direction of metabolic flux to the C19 pathway by catalyzing 22-OH-BNC-CoA to form 22-oxo-BNC-CoA. So, herein, we constructed a strain named MFKDΔ*hsd4A* by knocking out *hsd4A* in MFKD. As predicted, the HPLC elution profiles showed that metabolic flow was significantly twisted, as shown in Figure 3c. A total of 81.47% 9-OHBA was accumulated compared with 4.13% in the strain MFKD, which confirmed the predicted function of *hsd4A* (Table 1). Moreover, the selectivity of 9-OHAD reduced to 4.07% from 78.09%. A concentration of 6.16 g 9-OHBA per liter was reached from 10 g/L phytosterols, and the molar yield was 72.74%.

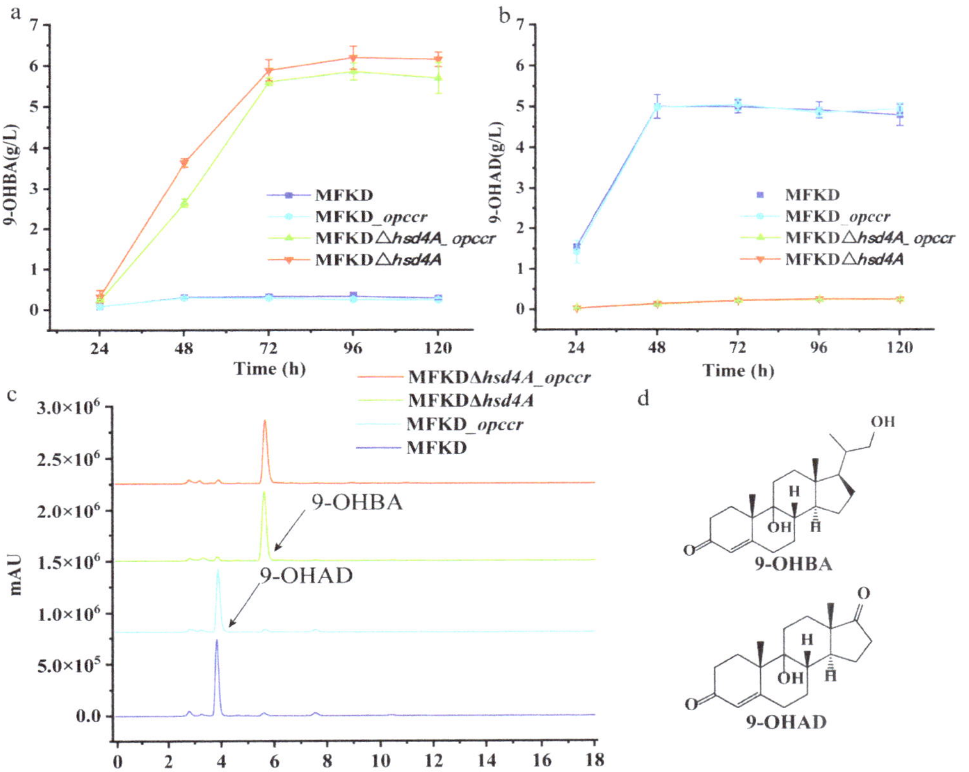

Figure 3. Phenotypic analyses of the metabolites of phytosterol produced by MFKD and its mutants. (**a**,**b**) Time course of 9-OHBA and 9-OHAD accumulation in mutant strains MFKD, MFKD_*opccR*, MFKDΔ*hsd4A*, and MFKDΔ*hsd4A_opccR*; (**c**) high-performance liquid chromatography (HPLC) elution profiles of metabolites produced via phytosterol bioconversion by MFKD, MFKD_*opccR*, MFKDΔ*hsd4A*, and MFKDΔ*hsd4A_opccR*; (**d**) structure of 9-OHBA and 9-OHAD. Error bars represent standard deviations.

Table 1. The relative selectivity of the steroid intermediates produced by MFKD and its mutants.

Strain	Relative Selectivity (%)			
	9-OHAD	9-OHBA	AD	BA
MFKD	78.09 ± 0.21	4.21 ± 0.08	0.82 ± 0.01	0.09 ± 0.01
MFKD_*opccR*	77.98 ± 2.06	4.76 ± 0.31	0.87 ± 0.16	0.12 ± 0.03
MFKDΔ*hsd4A*	4.07 ± 0.12	81.47 ± 0.04	0.35 ± 0.01	0.21 ± 0.01
MFKDΔ*hsd4A*_ *opccR*	4,67 ± 0.13	81.22 ± 0.37	0.63 ± 0.22	0.19 ± 0.01
MFKDΔ*hsd4A*Δ*fadA5*	0.90 ± 0.08	95.13 ± 0.46	0.37 ± 0.08	0.19 ± 0.02

As shown in Figure 1, reductase OpccR has activities in both 3-OPC-CoA and 3-OPA according to Peng's study [5]. It turned out that Peng's and Song's studies came to opposite conclusions regarding the role of *mnopccR* in *Mycobacterium neoaurum* [5,20]. Enlighted by their research, in *M. fortuitum* ATCC 35855, we also found an *mnopccr* (designated *opccR* in this study) that shared 76.3% identity with *mnopccr* in *M. neoaurum* CCTCC AB2019054 using amino acid BLAST. To test the function of OpccR in *M. fortuitum* ATCC 35855, the strains MFKD_*opccR* and MFKDΔ*hsd4A_opccR* were constructed by overexpressing *opccR*. However, after the inducement of phytosterols, there were no significant differences between the mutant strains and original strains, as shown in Figure 3. The selectivity of 9-OHBA slightly increased to 4.76% in MFKD_*opccR* from 4.21% in MFKD by 13.06%, and 9-OHAD reduced to 77.98% from 78.09% (Table 1).

These data indicated that the activity of the OpccR enzyme was much weaker than that in *Mycobacterium neoaurum*. In conclusion, the mutant strain MFKDΔ*hsd4A* was a good 9-OHBA-producing strain, although the by-product 9-OHAD was still present.

2.3. Deletion of fadA5 Improves the Selectivity of 9-OHBA by Eliminating 9-OHAD

The accumulation of 9-OHAD in the MFKDΔ*hsd4A* strain suggested that the C19 pathway of phytosterol degradation was not completely blocked. In order to further enhance the proportion of 9-OHBA and remove impurities, we deleted *fadA5* in MFKDΔ*hsd4A* to construct MFKDΔ*hsd4A*Δ*fadA5*. In line with our expectation, the double mutant MFKDΔ*hsd4A*Δ*fadA5* reduced 9-OHAD from 4.19% to 0.90%. The proportion of 9-OHBA reached 95.13%, which is an improvement of 13.62% compared with the former strain. The molar yield of 9-OHBA reached 80.33%. A concentration of 6.81 g 9-OHBA per liter was reached in MFKDΔ*hsd4A*Δ*fadA5* after it was cultured with 10 g/L phytosterols, as shown in Figure 4.

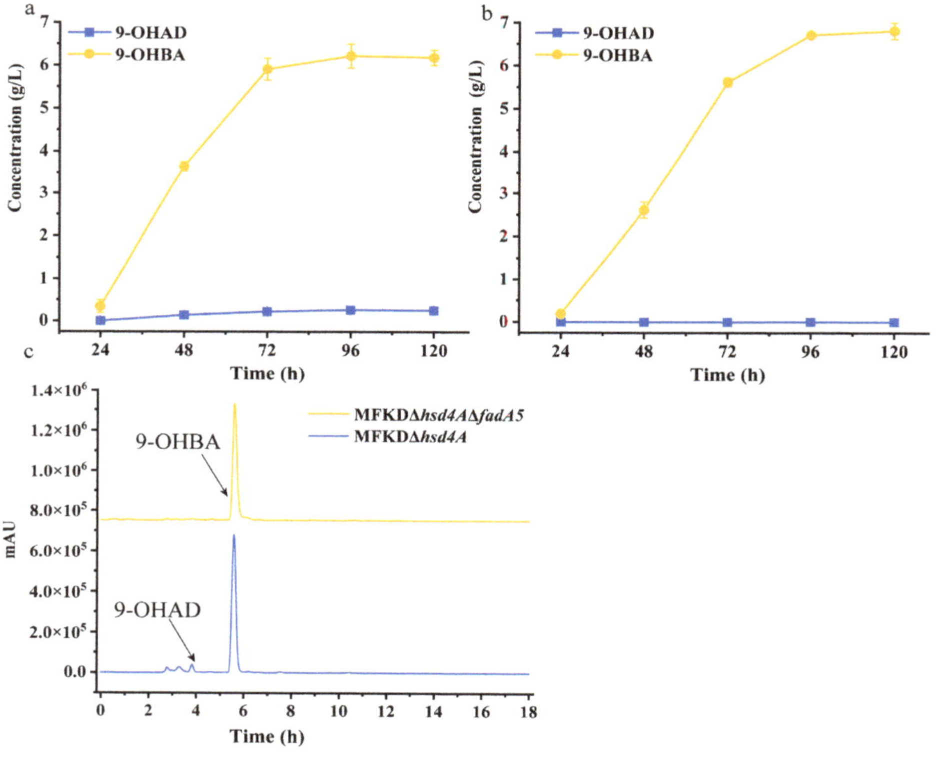

Figure 4. Phenotypic analyses of the metabolites of phytosterol produced by the 9-OHBA-producing strain. (**a**,**b**) Time profiles of product accumulation of mutant strains MFKDΔ*hsd4A* and MFKDΔ*hsd4A*Δ*fadA5*, respectively; (**c**) high-performance liquid chromatography (HPLC) elution profiles of metabolites produced via phytosterol bioconversion by MFKDΔ*hsd4A* and MFKDΔ*hsd4A*Δ*fadA5*. Error bars represent standard deviations.

2.4. Evaluation of the 9-OHBA-Producing Strain

As 9-OHAD production was almost eliminated in the double mutant MFKDΔ*hsd4A*Δ*fadA5*, we performed high-concentration fermentation using this strain. To assess the potential of the 9-OHBA-producing strain for transforming phytosterols into 9-OHBA, the strain MFKDΔ*hsd4A*Δ*fadA5* was cultured in the fermentation medium with 10, 15, and 20 g/L phytosterols.

As shown in Figure 5, we obtained 6.81 g/L, 9.28 g/L, and 7.76 g/L 9-OHBA from 10, 15, and 20 g/L phytosterols, respectively. The molar yields of 9-OHBA were 80.33%, 73.07%, and 45.82%, respectively. The selectivity of the main 9-OHBA product decreased to 90.98% when the concentration of phytosterols reached 15 g/L and was 86.92% when cultured with 20 g/L phytosterols. Meanwhile, the by-product of 9-OHAD increased to 3.80% when the concentration of phytosterols reached 20 g/L.

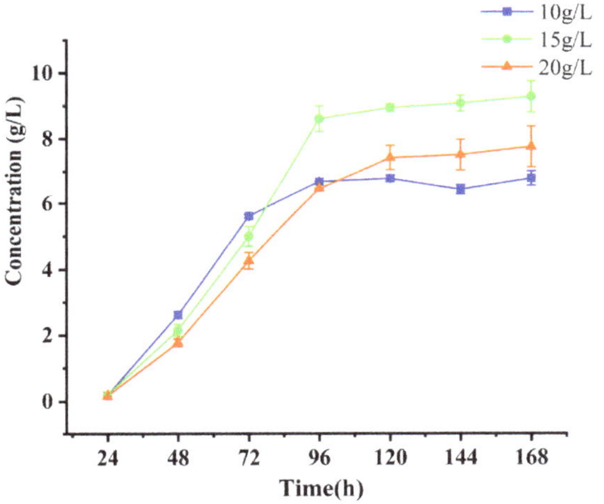

Figure 5. Time profiles of 9-OHBA accumulation by MFKDΔ*hsd4A*Δ*fadA5* with different concentrations of phytosterols. Error bars represent standard deviations.

3. Discussion

Mycobacterium fortuitum ATCC 35855 is a fast-growing strain and an outstanding producer of 9-OHAD. The mutant strain MFKD knocked out of five putative *kstDs* in our previous work prevented the degradation of the main nucleus to a large extent and finally accumulated 9-OHAD primarily and 9-OHBA as a by-product [19]. Owing to the extensive and in-depth research ons genes related to the C19 and C22 pathway in mycobacterial strains and the irreplaceable role of 9-OHBA in the synthesis of corticosteroids, we chose MFKD as a starting strain to obtain 9-OHBA and to find out the functions of related genes in *Mycobacterium fortuitum*.

Hsd4A, characterized as an enzyme that catalyzes 22-OH-BNC-CoA to form 22-oxo-BNC-CoA, is always chosen to regulate metabolic flux to produce C19 and C22 steroid intermediates [17]. In MFKD, we found one putative gene, *hsd4A*, whose product's amino acids shared an identity of 78.8% with Hsd4A in *M. neoaurum* DSM 44074, which has been reported to have catalytic activities, indicating that it is a putative dehydrogenase [18]. Therefore, we knocked out *hsd4A* in the mutant strain MFKD, and the direction of metabolic flux twisted to the C22 pathway as expected. The mutant strain MFKDΔ*hsd4A* accumulated 9-OHBA as the main product with a selectivity of 81.47%. However, 4.07% of the by-product of 9-OHAD still remained, indicating that isozymes of Hsd4A still exist. MnOpccr was first identified in *M. neoaurum* CCTCC AB2019054 as a bifunctional enzyme that catalyzes both the 4-e reduction of 3-OPC-CoA by the C terminal domain and the 2-e reduction of 3-OPA to form 4-BA by the N terminal domain [5]. The research also found that the inactivation of *mnOpccr* can eliminate BA, while the overexpression of *mnOpccr* with *hsd4A* inactivation can result in the sole production of BA from phytosterols [5]. Therefore, in order to improve metabolic flux in the C22 steroid pathway and eliminate impurities, we decided to enhance the expression of OpccR.

Given that OpccR has catalytic activities in both 3-OPC-CoA and 3-OPA, we constructed the mutant strains MFKD_*opccR* and MFKDΔ*hsd4A_opccR*. However, in this study, its overexpression neither in the MFKD strain nor in the mutant strain MFKDΔ*hsd4A* resulted in an evident increase in 9-OHBA. Specifically, the selectivity of 9-OHBA slightly increased by 13.06% to 4.76% in MFKD_*opccR* from 4.21% in MFKD. Given that the N and C terminal domains of OpccR are in charge of different reactions, OpccR in this MFKD may have a substrate preference for 3-OPC-CoA over 3-OPA. Another example is in *M. neoaurum* HGMS2. The OpccR enzyme was found to inhibit the accumulation of BA in HGMS2 [20]. This result contradicted Peng's report, suggesting a different metabolic pathway. The limitation in catalytic activity and relatively low protein expression level of OpccR also contribute to this phenomenon. In addition, other isoenzymes may be present in this strain, the functions of which require further investigation.

FadA5 has been reported to cleave 3,22-dioxo-chol-4-ene-24-oyl-CoA to yield 3-OPC-CoA and Ac-CoA, and prefers the steroid CoA substrate [12]. In MFKD, we identified a putative gene, *fadA5*, which coded the thiolase FadA5. The amino acids of FadA5 shared an identity of 88.4% with *M. neoaurum* DSM 44074. As shown in Figure 1, we speculated that the knock-out of *fadA5* would further remove the by-product 9-OHAD. Hence, we inactivated FadA5 from MFKDΔ*hsd4A* to construct MFKDΔ*hsd4A*Δ*fadA5*. As expected, the selectivity of 9-OHAD reduced from 4.07% to 0.90%. Meanwhile, the target product 9-OHBA reached 95.13% from 81.47%, suggesting that the C19 pathway was blocked to a large extent. Finally, we obtained 6.81 g/L 9-OHBA after culturing with 10 g/L phytosterols for 144h using the mutant strain MFKDΔ*hsd4A*Δ*fadA5*. This yield ranks highest among the reported strains, compared with 3.58 g/L 9-OHBA from 5 g/L phytosterols and 2.73 g/L 9-OHBA from 10 g/L phytosterols from Yuan's article [18]. The remaining 9-OHAD suggests that isozymes of Hsd4A and FadA5 still exist.

Finally, we obtained 6.81 g/L 9-OHBA with a selectivity of 95.13% and a molar yield of 80.33% after culturing with 10 g/L phytosterols for 144 h using the mutant strain MFKDΔ*hsd4A*Δ*fadA5*. This yield and selectivity rank highest among the strains reported so far using the shake flask fermentation method. Considering the best performance of the mutant strain MFKDΔ*hsd4A*Δ*fadA5* in producing 9-OHBA, we further evaluated its capacity under high concentrations of phytosterols. The concentration of 9-OHBA reached 9.28 g/L, which was highest when the strain MFKDΔ*hsd4A*Δ*fadA5* was cultured with 15 g/L phytosterols. However, the molar yield of 9-OHBA reduced to 73.07%. Only 7.76 g/L 9-OHBA was obtained when cultured with 20 g/L phytosterols, with a poor molar yield of 45.82%. Several reasons could be responsible for this phenomenon. Firstly, due to the low water solubility of phytosterols, their dispersion in aqueous phase is poor, thus limiting the accessibility of mycobacteria cells during phytosterol biotransformation [23]. The dispersibility of phytosterols further deteriorates with an increase in concentration. Therefore, dispersants such as surfactants, CDs (cyclic oligosaccharides), and water-miscible organic solvents are added to increase the solubility of the hydrophobic sterols in aqueous media. The ADD production of *M. neoaurum* VKPM Ac-1656 increased using the surfactant Tween-80 and the over-crosslinked polystyrene resin MN-200. Other surfactants like lecithin, polyoxyethylene (10) nonylphenyl ether (TX-40), and sucrose ester DK-Ester P-160 also proved to promote the transformation of phytosterols [24–26]. In addition, phytosterols and their derivatives inhibited cell growth by reducing the utilization of carbon sources and further hindered bioconversion [27]. Also, this effect became more prominent with an increase in phytosterol concentration. Poor fermentation conditions and low cell concentration in the shake flask also inhibited conversion efficiency. Large fermenters and high-oxygen conditions can alleviate this situation accordingly.

4. Materials and Methods

4.1. Bacterial Strains, Plasmids, Reagents, and Culture Conditions

The *Mycobacterium* strains and plasmids used in this study are listed in Table 2. The mutant *Mycobacterium* strains were constructed based on the strain MKFD [19], a KstD-deficiency strain derived from *Mycobacterium fortuitum* ATCC 35855 maintained in our laboratory. We employed the homologous recombinant knockout plasmid pKADel (obtained by combining p2NIL and pGoal19) and the P40 gene integrative vector, which was derived from the plasmid pMV306, for the gene modification of *M. fortuitum* [19].

LBT medium (10.0 g/L NaCl, 10.0 g/L tryptone, 5.0 g/L yeast extract, and 2.0 g/L Tween-80 (pH 7.0)) was used for aerobic cultivation of *M. fortuitum* at 30 °C and 200 rpm. MT medium, which contained 20 g/L glucose, 12 g/L $(NH4)_2HPO_4$, 0.5 g/L $MgSO_4$, 0.5 g/L $NaNO_3$, 3 g/L citric acid, 0.05 g/L ammonium ferric citrate, and 0.2% Tween-80 (v/v), was used as a fermentation medium by *Mycobacterium* cells. A total of 3mL seed medium was transferred to 30 mL of MT medium in a 250 mL shaker flask with a baffle when the optical density reached the mid-log exponential phase. Shaker flasks with baffles were used to increase the dissolved oxygen level in the culture medium effectively and enhance the contact between cells and air. The initial pH was adjusted to 7.5 and the fermentation was carried out at 30 °C and 220 rpm. The phytosterols consisted of 45% β-sitosterol, 37% campesterol, and 18% stigmasterol, purchased from Yunnan Biological Products Co., Ltd. (Kunming, China). A total of 60 g/L phytosterol mother liquor was configured by mixing phytosterol and (2-hydroxypropyl)-β-cyclodextrin (HP-β-CD) in water with a ratio of 1:3 (m/m), stirring for 15 min, ultrasound dispersing for 20 min, and repeating three times. AD, ADD, BA, 1,4-BA, 9-OHBA, and 9-OHAD were purchased from Sigma-Aldrich (Shanghai, China); all other reagents used were of analytical grade or higher unless noted otherwise.

4.2. Bioinformatic Analysis

The genome of ATCC 35855 has been sequenced previously and deposited in the GenBank database under the accession number CP110127 [19]. The putative genes for *hsd4A*, *fadA5*, and *opccR* were identified by comparison with known gene sequences from the NCBI database. Amino acid sequences of putative Hsd4A, FadA5, and OpccR were aligned by ClustalW; phylogenetic trees were constructed by neighbor-joining algorithm using MEGA 11 software.

4.3. Construction of Mutant Strains

Gene deletion and overexpression strategies were used to construct recombinant MFKD mutants. PKADel, obtained by combining p2NIL and pGoal19 [19,28], was used as the knockout plasmid via homologous recombination (Figure S1). Specifically, we amplified DNA sequences 1.2kb in length, located upstream and downstream of the target gene from the MFKD genome, including *hsd4A* and *fadA5* (Table S1). Then, the fragments were joined into linearized pKADel digested with AflII and SalI, constructing pKADelΔ*hsd4A* and pKADelΔ*fadA5*. Afterward, we electroporated the knockout plasmid into competent mycobacterial cells following previous procedures [19]. Finally, we screened the positive colonies using primers located upstream and downstream. The colonies with shorter fragments were confirmed to be positive. The vector p40 (pMV306 with the Psmyc promoter) was used to overexpress related genes in the mutant strains. *Opccr* was amplified from *M. fortuitum* ATCC 35855, and then, inserted into linearized P40 digested with AflII and HindIII to construct p40-*opccR*. Next, the plasmid was electroporated into the same competent mycobacterial cells as before.

Table 2. Strains and plasmids used in this study.

Name	Description	Source
Strains		
Escherichia coli	*E. coli* DH5α	Vazyme Biotech Co., Ltd., Nanjing, China
MFKD	9-OHAD producer, *kstD1&2&3&4&5* deletion mutant of ATCC 35855	Our lab
MFKD_*opccR*	ATCC 35855 *opccR* overexpression in MFKD via p40-*opccR*	This study
MFKDΔ*hsd4A*	*hsd4A* deletion mutant of MFKD	This study
MFKDΔ*hsd4A*_ *opccR*	ATCC 35855 *opccR* overexpression in MFKDΔ*hsd4A* via p40-*opccR*	This study
MFKDΔ*hsd4A*Δ*fadA5*	*hsd4A* and *fadA5* double-deletion mutant of MFKD	This study
Plasmids		
pKADel	Plasmid for allelic exchange, Pag85-lacZ Phsp60-sacB, AprR, KanR	[19]
pKADelΔ*hsd4A*	pKADel carrying two homologous arms of *hsd4A*	This study
pKADelΔ*fadA5*	pKADel carrying two homologous arms of *fadA5*	This study
p40	pMV306 with Psmyc promoter, KanR	[19]
p40-*opccR*	p40 possessing *opccR* from *M. fortuitum* ATCC 35855	This study

4.4. Bioconversion and Analytical Methods

The mutant strains were first cultured in the LBT medium to the mid-logarithmic growth stage, and then, transferred to the MT medium containing phytosterol at 10% (v/v). The initial concentration of phytosterol was 10 g/L. Later, a concentration gradient was tested to further determine the strains' ability for phytosterol bioconversion. An appropriate amount of evenly dispersed phytosterol mother liquor was taken and diluted with MT medium to the desired concentration to obtain the different fermentation concentrations of phytosterol.

In the *M. fortuitum* mutant fermentation experiments, samples were taken every 24 h for 5–7 days, and three replicates were used to measure and quantify steroids. Culture samples (0.5 mL) were extracted on a vortex mixer with 1 mL of ethyl acetate for 20 min before centrifugation at 12,000× g for 1 min. For HPLC, the organic phase of the sample was redissolved in methanol after volatilizing, and then, filtered through a 0.22 μm microporous membrane. Separation was performed on an Agilent XDB-C18 column (4.6 × 250 mm; 40 °C), and a UV/visible detector (254 nm) was employed to detect the steroid substrate conversion rates with methanol/water (80:20, v/v). The flow rate was 0.8 mL/min.

The molar yield (*My*) of steroid products 9-OHAD and 9-OHBA was calculated using the following equation:

$$My = \frac{Ma}{Mt} \times 100\%,$$

where *Ma* and *Mt* are the moles of actual steroid products and theoretical steroid products, respectively.

5. Conclusions

In this study, after knocking out *hsd4A* and *fadA5*, we successfully constructed an ideal 9-OHBA producer, MFKDΔ*hsd4A*Δ*fadA5*, with high yield and selectivity. This result proves that Hsd4A and FadA5 are crucial enzymes controlling the C19 metabolic pathway of steroids in *Mycobacterium fortuitum* ATCC 35855. But the function of OpccR in *M. fortuitum*

ATCC 35855 still requires further research. Our work provides new insights into the strategies and methods for the production of relevant steroid intermediates.

Supplementary Materials: The following supporting information can be downloaded at: https://www.mdpi.com/article/10.3390/ijms25073579/s1.

Author Contributions: S.H.: conceptualization, data curation, formal analysis, visualization, roles/writing—original draft; X.L.: investigation, supervision, project administration; X.Z. and B.H.: methodology, software; W.L., H.W. and C.Y.: resources; Y.L.: writing—review and editing; B.Z.: funding acquisition, validation, writing—review and editing. All authors have read and agreed to the published version of the manuscript.

Funding: This work was supported by the National Key R&D Program of China (No. 2017YFE0112700) through the "Science and technology innovation" team project of Hubei province in 2022 and "unveiling system" technology project of Hubei province in 2022.

Institutional Review Board Statement: Not applicable.

Informed Consent Statement: Not applicable.

Data Availability Statement: The genome sequencing information of *M. fortuitum* ATCC 35855 has been deposited in the GenBank database with accession number CP110127. The accession numbers of the *hsd4A*, *fadA5*, and *opccR* gene sequences are OP729274.1, WP_019512519.1, and OP729275.1, respectively.

Acknowledgments: We sincerely thank T. Parish (Department of Infectious and Tropical Diseases, United Kingdom) for providing the plasmids, p2NIL and pGOAL19, and W. R. Jacobs (Howard Hughes Medical Institute) for providing the plasmid pMV306.

Conflicts of Interest: The authors declare no conflicts of interest. The funder had no role in the design or performance of the study.

References

1. Donova, M.V. Steroid Bioconversions. *Methods Mol. Biol.* **2017**, *1645*, 1–13. [CrossRef] [PubMed]
2. Rugutt, J.K.; Rugutt, K.J. Antimycobacterial activity of steroids, long-chain alcohols and lytic peptides. *Nat. Prod. Res.* **2012**, *26*, 1004–1011. [CrossRef] [PubMed]
3. Seth, D.; Kamat, D. Intranasal Steroid Therapy for Allergic Rhinitis. *Pediatr. Ann.* **2019**, *48*, e43–e48. [CrossRef] [PubMed]
4. Liu, Y.J.; Ji, W.T.; Song, L.; Tao, X.Y.; Zhao, M.; Gao, B.; Meng, H.; Wang, F.Q.; Wei, D.Z. Transformation of phytosterols into pregnatetraenedione by a combined microbial and chemical process. *Green Chem.* **2022**, *24*, 3759–3771. [CrossRef]
5. Peng, H.; Wang, Y.; Jiang, K.; Chen, X.; Zhang, W.; Zhang, Y.; Deng, Z.; Qu, X. A Dual Role Reductase from Phytosterols Catabolism Enables the Efficient Production of Valuable Steroid Precursors. *Angew. Chem. Int. Ed. Engl.* **2021**, *60*, 5414–5420. [CrossRef] [PubMed]
6. Zhao, A.; Zhang, X.; Li, Y.; Wang, Z.; Lv, Y.; Liu, J.; Alam, M.A.; Xiong, W.; Xu, J. *Mycolicibacterium* cell factory for the production of steroid-based drug intermediates. *Biotechnol. Adv.* **2021**, *53*, 107860. [CrossRef] [PubMed]
7. Donova, M.V. Transformation of steroids by actinobacteria: A review. *Appl. Biochem. Microbiol.* **2007**, *43*, 1–14. [CrossRef]
8. Almeida, C.A.S.; Baggio, S.R.; Mariutti, L.R.B.; Bragagnolo, N. One-step rapid extraction of phytosterols from vegetable oils. *Food Res. Int.* **2020**, *130*, 108891. [CrossRef]
9. Andryushina, V.A.; Rodina, N.V.; Stytsenko, T.S.; Huy, L.D.; Druzhinina, A.V.; Yaderetz, V.V.; Voishvillo, N.E. Conversion of soybean sterols into 3,17-diketosteroids using actinobacteria *Mycobacterium neoaurum*, *Pimelobacter simplex*, and *Rhodococcus erythropolis*. *Appl. Biochem. Microbiol.* **2011**, *47*, 270–273. [CrossRef]
10. Fernandez-Cabezon, L.; Galan, B.; Garcia, J.L. New Insights on Steroid Biotechnology. *Front. Microbiol.* **2018**, *9*, 958. [CrossRef]
11. McLeod, M.P.; Warren, R.L.; Hsiao, W.W.L.; Araki, N.; Myhre, M.; Fernandes, C.; Miyazawa, D.; Wong, W.; Lillquist, A.L.; Wang, D.; et al. The complete genome of *Rhodococcus* sp RHA1 provides insights into a catabolic powerhouse. *Proc. Natl. Acad. Sci. USA* **2006**, *103*, 15582–15587. [CrossRef]
12. Schaefer, C.M.; Lu, R.; Nesbitt, N.M.; Schiebel, J.; Sampson, N.S.; Kisker, C. FadA5 a thiolase from *Mycobacterium tuberculosis*: A steroid-binding pocket reveals the potential for drug development against tuberculosis. *Structure* **2015**, *23*, 21–33. [CrossRef] [PubMed]
13. Wipperman, M.F.; Sampson, N.S.; Thomas, S.T. Pathogen roid rage: Cholesterol utilization by *Mycobacterium tuberculosis*. *Crit. Rev. Biochem. Mol.* **2014**, *49*, 269–293. [CrossRef] [PubMed]
14. Yao, K.; Wang, F.Q.; Zhang, H.C.; Wei, D.Z. Identification and engineering of cholesterol oxidases involved in the initial step of sterols catabolism in *Mycobacterium neoaurum*. *Metab. Eng.* **2013**, *15*, 75–87. [CrossRef] [PubMed]

15. Donova, M.V.; Egorova, O.V. Microbial steroid transformations: Current state and prospects. *Appl. Microbiol. Biotechnol.* **2012**, *94*, 1423–1447. [CrossRef] [PubMed]
16. Liu, X.C.; He, B.R.; Zhang, J.X.; Yuan, C.Y.; Han, S.W.; Du, G.L.; Shi, J.P.; Sun, J.S.; Zhang, B.G. Phytosterol conversion into C9 non-hydroxylated derivatives through gene regulation in. *Appl. Microbiol. Biotechnol.* **2023**, *107*, 7635–7646. [CrossRef]
17. Xu, L.Q.; Liu, Y.J.; Yao, K.; Liu, H.H.; Tao, X.Y.; Wang, F.Q.; Wei, D.Z. Unraveling and engineering the production of 23,24-bisnorcholenic steroids in sterol metabolism. *Sci. Rep.* **2016**, *6*, 21928. [CrossRef]
18. Yuan, C.Y.; Ma, Z.G.; Zhang, J.X.; Liu, X.C.; Du, G.L.; Sun, J.S.; Shi, J.P.; Zhang, B.G. Production of 9,21-dihydroxy-20-methyl-pregna-4-en-3-one from phytosterols in *Mycobacterium neoaurum* by modifying multiple genes and improving the intracellular environment. *Microb. Cell Fact.* **2021**, *20*, 229. [CrossRef]
19. Liu, X.; Zhang, J.; Yuan, C.; Du, G.; Han, S.; Shi, J.; Sun, J.; Zhang, B. Improving the production of 9alpha-hydroxy-4-androstene-3,17-dione from phytosterols by 3-ketosteroid-Delta(1)-dehydrogenase deletions and multiple genetic modifications in *Mycobacterium fortuitum*. *Microb. Cell Fact.* **2023**, *22*, 53. [CrossRef]
20. Song, S.; He, J.; Gao, M.; Huang, Y.; Cheng, X.; Su, Z. Loop pathways are responsible for tuning the accumulation of C19- and C22-sterol intermediates in the mycobacterial phytosterol degradation pathway. *Microb. Cell Fact.* **2023**, *22*, 19. [CrossRef]
21. Zhang, J.X.; Zhang, R.J.; Song, S.K.; Su, Z.D.; Shi, J.P.; Cao, H.J.; Zhang, B.G. Whole-Genome Analysis of *Mycobacterium neoaurum* DSM 1381 and the Validation of Two Key Enzymes Affecting C22 Steroid Intermediates in Sterol Metabolism. *Int. J. Mol. Sci.* **2023**, *24*, 6148. [CrossRef] [PubMed]
22. Lu, R.; Schaefer, C.M.; Neshitt, N.M.; Kuper, J.; Kisker, C.; Sampson, N.S. Catabolism of the Cholesterol Side Chain in *Mycobacterium tuberculosis* Is Controlled by a Redox-Sensitive Thiol Switch. *ACS Infect. Dis.* **2017**, *3*, 666–675. [CrossRef]
23. Turk, M.; Lietzow, R. Stabilized nanoparticles of phytosterol by rapid expansion from supercritical solution into aqueous solution. *AAPS PharmSciTech* **2004**, *5*, e56. [CrossRef] [PubMed]
24. Mancilla, R.A.; Little, C.; Amoroso, A. Efficient Bioconversion of High Concentration Phytosterol Microdispersion to 4-Androstene-3,17-Dione (AD) by *Mycobacterium* sp B3805. *Appl. Biochem. Biotechnol.* **2018**, *185*, 494–506. [CrossRef] [PubMed]
25. Wang, Z.F.; Huang, Y.L.; Rathman, J.F.; Yang, S.T. Lecithin-enhanced biotransforation of cholesterol to androsta-1,4-diene-3,17-dione and androsta-4-ene-3,1 7-dione. *J. Chem. Technol. Biotechnol.* **2002**, *77*, 1349–1357. [CrossRef]
26. Zhou, L.F.; Li, H.; Xu, Y.A.; Liu, W.; Zhang, X.M.; Gong, J.S.; Xu, Z.H.; Shi, J.S. Effects of a nonionic surfactant TX-40 on 9 alpha-hydroxyandrost-4-ene-3,17-dione biosynthesis and physiological properties of *Mycobacterium* sp. LY-1. *Process Biochem.* **2019**, *87*, 89–94. [CrossRef]
27. Su, L.Q.; Xu, S.P.; Shen, Y.B.; Xia, M.L.; Ren, X.X.; Wang, L.F.; Shang, Z.H.; Wang, M. The Sterol Carrier Hydroxypropyl-beta-Cyclodextrin Enhances the Metabolism of Phytosterols by *Mycobacterium neoaurum*. *Appl. Environ. Microbiol.* **2020**, *86*, e00441-20. [CrossRef]
28. Parish, T. Use of a flexible cassette method to generate a double unmarked *Mycobacterium tuberculosis* tlyA plcABC mutant by gene replacement. *Microbiology* **2000**, *146*, 1969–1975. [CrossRef]

International Journal of
Molecular Sciences

MDPI

Article

Inactivation of *sacB* Gene Allows Higher 2,3-Butanediol Production by *Bacillus licheniformis* from Inulin

Emanoel Gergov [1], Penka Petrova [1], Alexander Arsov [1], Ina Ignatova [2], Lidia Tsigoriyna [2], Nadya Armenova [2] and Kaloyan Petrov [2,*]

[1] Institute of Microbiology, Bulgarian Academy of Sciences, 1113 Sofia, Bulgaria; emanoelgergov@microbio.bas.bg (E.G.); ppetrova@microbio.bas.bg (P.P.); al.arsov@microbio.bas.bg (A.A.)

[2] Institute of Chemical Engineering, Bulgarian Academy of Sciences, 1113 Sofia, Bulgaria; ina.ignatova@iche.bas.bg (I.I.); lidinka29@gmail.com (L.T.); nadq.armenova@gmail.com (N.A.)

* Correspondence: kkpetrov@iche.bas.bg

Abstract: *Bacillus licheniformis* 24 (BL24) is an efficient, non-pathogenic producer of 2,3-butanediol (2,3-BD). However, during inulin fermentation, the strain produces large amounts of exopolysaccharides (EPS), which interfere with the process' performance. The present study aims to investigate the effect that inactivation of the *sacB* gene, encoding levansucrase in BL24, has on 2,3-BD production efficiency. Knockout of the *sacB* gene was accomplished via insertional inactivation. The *sacB*-knockout variant formed 0.57 g/L EPS from sucrose and 0.7–0.8 g/L EPS from glucose and fructose, a 15- and 2.5-fold reduction relative to the wild type, respectively. Likewise, during batch fermentation with soluble inulin Frutafit® CLR, the mutant BLΔ*sacB* produced significantly less EPS than the wild type, allowing the maintenance of pH at values favoring 2,3-BD synthesis. At pH 6.50, BLΔ*sacB* reached a record titer of 128.7 g/L 2,3-BD, with productivity of 1.65 g/L/h, and a yield of 85.8% of the theoretical maximum. The obtained concentration of 2,3-BD is two-fold higher compared to that of the wild type. Subsequent RT-qPCR assays confirmed a successful *sacB* knockout. Three of the genes involved in inulin hydrolysis (*sacA*, *sacC*, and *fruA*) maintained their expression levels compared to the wild type, while that of *levB* increased. Although total EPS accumulation could not be completely eliminated via *sacB* gene knockout alone, the overall reduction in EPS content has enabled the highest yield of 2,3-BD from inulin to date, a promising result for the industrial production from inulin-rich substrates.

Keywords: 2,3-butanediol; *Bacillus licheniformis*; *sacB*; exopolysaccharides; inulin

1. Introduction

Bacillus licheniformis is an unpretentious soil bacterium with wide and varied industrial applications. As a Gram-positive, spore-forming, and non-pathogenic species, *B. licheniformis* attracted great biotechnological interest as a microbial factory for the synthesis of enzymes (protease, keratinase, amylase, cellulase, allantoinase, chitinase, arabinase, levanase, etc.), bioactive compounds (lipopeptides), exopolysaccharides (EPS), polyhydroxyalkanoates, and hydrogen [1–4]. Particularly important, however, is the application of the species in obtaining 2,3-butanediol (2,3-BD), a greatly demanded platform chemical for butadiene, methyl ethyl ketone, and polyurethane maleamide production, often used as an antifreeze agent and fuel additive [5,6]. Currently, 2,3-BD is produced by chemical synthesis from petroleum. The microbial production of 2,3-BD is considered white biotechnology as it replaces the chemical synthesis of an industrially important reagent with a microbial one, with the possibility of utilizing renewable substrates, including cellulose and lignocellulosic materials, molasses, food waste, and inulin [7–12]. Due to rising oil prices, pilot plants were built in the United States and China to process pretreated lignocellulosic waste into 2,3-BD. Despite recent scientific progress, industrial microbial production of 2,3-BD has yet to be launched [13].

The highest productivity of 2,3-BD was traditionally reported in *Klebsiella pneumoniae* and *K. oxytoca* [14,15], both pathogenic bacterial species. However, considering the titer and yield from glucose, they are already equaled by GRAS (Generally Regarded As Safe) producers of the genera *Penibacillus* and *Bacillus* [16,17]. Among them, *B. licheniformis* is preferred, as it can achieve a high concentration of 2,3-BD (up to 150 g/L) and a yield close to the theoretical maximum (0.44–0.47 g/g), both from glucose and fructose [13,18]. Furthermore, this species is also capable of converting inulin-containing feedstocks, which is promising for the profitability of the potential industrial process [19].

Inulin is a poly-fructan, abundant in some non-food plants, and is therefore a preferred substrate for use in microbial fermentations. It is a reserve polysaccharide in the *Asteraceae* family, which includes chicory, dahlia, and Jerusalem artichoke (JAT), whichgrow in dry and infertile soils, and are known as C4-plants that are capable of active carbon dioxide fixation [20,21]. Derived from wild and adaptable plants, inulin flours or extracts are relatively cheap. For example, the retail price range for China chicory is between USD 0.38 and USD 0.76 per kilogram [22]. Therefore, inulin-containing materials are considered a cheap and abundant renewable feedstock for bioprocessing. They are already widely used in the biotechnological industrial production of inulinases [23,24], fructose and high fructose syrup [25,26], prebiotic fructooligosaccharides, single-cell protein, single-cell oil, organic acids such as citric, lactic, and gluconic acid, and bioethanol [27]. The bioprocesses for the production of acetone, butanol, and 2,3-butanediol are still under development [13,28].

However, the cost of microbial conversion to 2,3-BD increases with the necessary pretreatment steps, as all known biotechnologies to date carry out the inulin fermentation after enzymatic, acid, or combined acid-autoclaving hydrolysis to break down the inulin to fructose [29–32]. Currently, a significant amount of research concentrates on the efficient microbial production of 2,3-BD from inulin, but strains capable of simultaneous saccharification and fermentation (SSF) of inulin are rare [33–35].

Recently, we reported the isolation of such a strain, *B. lichenifomis* 24, which is a potent overproducer of 2,3-BD from soluble inulin without the need for preliminary hydrolysis. Although the strain did not degrade non-soluble Highly Dispersible inulin (Frutafit® HD), it produced at rates as high as 67.5 g/L 2,3-BD by direct fermentation of soluble inulin Frutafit® CLR [36]. However, increasingly large amounts of EPS formed when the processes were carried out at pH values above 6.00. Thus, the culture medium became thick and viscous, while the process at a higher pH than 6.25 was associated with ineffective aeration and the technical impossibility of conducting the fermentation. Consequently, the optimal pH for 2,3-BD production by *B. licheniformis* 24 (BL24) was never reached. When examining the acting enzymes, it was determined that at pH 6.25, the gene expression of *sacB* increased up to 197-fold compared to the process at pH 5.25 [36]. The enzyme sacB of *B. licheniformis* (E.C. 2.4.1.10, GH68) has dual hydrolase and transglycosylase activity. It is involved in the hydrolysis of inulin, but it is also responsible for the synthesis of levan in large quantities [37,38]. Therefore, levan synthesis can only be avoided by developing an engineered variant of the producer strain in which the levansucrase is blocked.

The present study aimed to knock out the *sacB* gene in BL24, investigate the effect of that modification on the total EPS synthesis and gene expression, and perform the process for maximal 2,3-BD production.

2. Results

2.1. Inactivation of sacB Gene in BL24

The purposeful inactivation of the *sacB* gene, encoding levan-producing levansucrase in BL24, was performed by insertional inactivation of a shuttle vector. Homologous recombination of a disrupted *sacB* and the reporter gene with the chromosome of BL24 was achieved in several cloning steps. The backbone of the construct was the integrative vector pBacTag-DYKDDDK (Figure 1). However, the selectable marker of the vector for erythromycin (EryR) appeared unsuitable because BL24 is resistant to erythromycin up to a concentration of 300 μg/mL. Since the host strain is susceptible to kanamycin at every concentration, aminoglycoside phosphotransferase from transposon Tn5, which confers resistance to kanamycin and neomycin (KanR/NeoR), was chosen as an appropriate reporter gene. Using the primer pairs Bac_F and Bac_R, the vector backbone was amplified with the excision of EryR and the T1T2T0 terminator sequence. The obtained partial version of pBacTag-DYKDDDK (4740 bp) was ligated by Gibson Assembly® to the fragment (1009 bp) which contained KanR with its promoter and which was previously PCR-amplified from the pCR®2.1-TOPO vector as a template. The resulting shuttle vector, pBac_KanR (5749 bp), was linearized with KpnI, amplified, and assembled with the truncated and mutated Δ*sacB* (685 bp) obtained as a PCR product with template BL24 chromosomal DNA. The final construct, pBac_KanR_Δ*sacB* (6432 bp), was used to transform BL24.

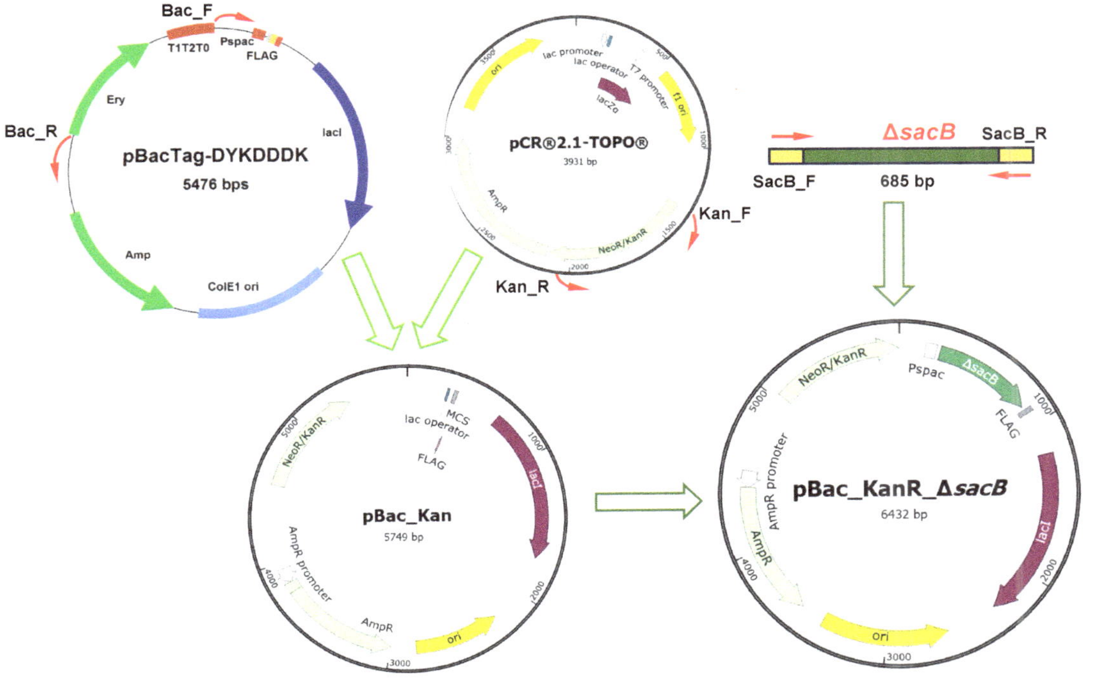

Figure 1. Cloning scheme. To eliminate the EryR gene and the adjacent terminator region, part of the pBac-TagDYKDDDK vector was amplified with a Bac_F and Bac_R primer pair, after which the resulting fragment (4738 bp) was ligated to KanR derived from the pCR®2.1-TOPO vector. The obtained construct pBac_Kan (5749 bp) was used for Gibson Assembly® with the truncated *sacB* gene of BL24, resulting in the final pBac_KanR_Δ*sacB*. The sites and the direction of amplification are indicated with red arrows.

After the transformation of electrocompetent *B. licheniformis* strain 24 and selection in the presence of 20 μg/mL kanamycin, 1.3×10^2 transformants were obtained. These were screened for levan production by cultivation in Luria–Bertani (LB) medium with 50 g/L sucrose, followed by extraction and measurement of the total EPS. The clones that did not form EPS from fructose were further analyzed. The presence of the integrated construct was confirmed by PCR targeting Δ*sacB* and the reporter kanamycin gene, as well as sequencing of the products.

2.2. EPS Synthesis by BLΔsacB from Sucrose and Monosaccharides

The total EPS production by the engineered BLΔ*sacB* was compared with those of BL24 during flask-batch fermentation without pH control in LB broth supplemented with 50 g/L sucrose, glucose, or fructose. In the medium containing sucrose, the wild type formed large amounts of EPS, reaching 8.78 g/L at the 30th hour of fermentation, while the *sacB*-knockout variant BLΔ*sacB* formed only 0.57 g/L, or a 15-fold lower amount (Figure 2a). From glucose and fructose, the wild type formed about 2 g/L of EPS, while BLΔ*sacB* formed between 0.7 and 0.8 g/L, or about 2.5 times less (Figure 2b).

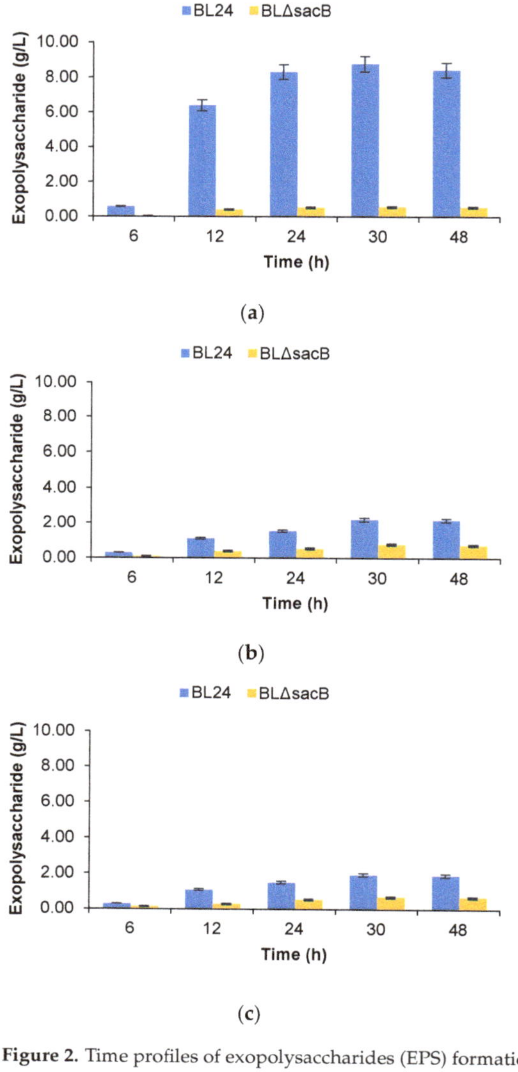

Figure 2. Time profiles of exopolysaccharides (EPS) formation by BL24 and its engineered variant, BLΔ*sacB*, during the course of flask-batch fermentation processes with a 50 g/L concentration of different sugars on a rotary shaker at 37 °C and 200 rpm for 48 h. (**a**) Sucrose; (**b**) Glucose; (**c**) Fructose.

2.3. Effect of pH on Inulin Conversion to 2,3-BD by the Engineered BLΔsacB

Among process parameters, pH maintenance is crucial for the successful conversion of inulin to 2,3-BD. With increasing pH, 2,3-BD increased; however, it increased alongside a high accumulation of EPS, leading to uncontrollable thickening of the culture broth. Thus, the highest pH at which inulin fermentation by native BL24 can occur is pH 6.25. In contrast, with the engineered variant BLΔ*sacB*, several processes at pH maintained at 6.25, 6.50, and 6.75 were carried out, as there was no significant production of EPS impeding the fermentation. Table 1 presents the obtained metabolites (2,3-BD, acetoin, and byproducts) at the moment of the highest production of 2,3-BD.

Table 1. Batch fermentation of soluble chicory flour Frutafit® CLR to 2,3-butanediol at different pH levels by BL24 and its engineered variant BLΔ*sacB*. The average final concentrations of the products obtained in three independent experiments are shown, with a standard deviation of 3%. All fermentations were conducted in a 1 L stirred fermenter at a temperature of 37.8 °C, aeration 3.68 vvm, and agitation 500 rpm. The fold change was calculated vs. 0 h at pH 5.25 for BL24 [36]. The comparative ΔΔCt method was used for the estimation of the relative abundance of each gene on the mRNA level. The 16S rRNA gene was used for normalization.

Strain	Conditions		Products					Maximum Gene Expression Levels				
	CFP [1] (g/L)	pH	2,3-BD (g/L)	Acetoin (g/L)	Glycerol (g/L)	LA [2] (g/L)	SA [3] (g/L)	*sacA* (FC [4])	*sacB* (FC [4])	*sacC* (FC [4])	*fruA* (FC [4])	*levB* (FC [4])
BL24 [5]	200	6.25	67.5 ± 3.5	0.3 ± 0.3	9.8 ± 1.4	ND	1.3 ± 0.1	66.26 ± 2.1	196.72 ± 6.1	163.14 ± 5.8	53.82 ± 2.4	0.40 ± 0.1
BLΔ*sacB*	200	6.25	75.5 ± 2.3	1.3 ± 1.2	14.1 ± 1.6	ND	0.5 ± 0.1	64.71 ± 2.3	ND	118.16 ± 5.4	49.13 ± 2.1	59.73 ± 1.9
BLΔ*sacB*	300	6.50	128.7 ± 4.1	1.8 ± 1.1	30.2 ± 2.5	ND	3.3 ± 0.2	68.45 ± 2.8	ND	126.64 ± 5.7	46.03 ± 2.3	82.16 ± 5.2
BLΔ*sacB*	300	6.75	96.4 ± 4.7	2.1 ± 2.0	10.9 ± 2.3	44.8 ± 2.8	2.0 ± 0.2	70.40 ± 3.2	ND	134.24 ± 5.8	48.79 ± 2.3	88.19 ± 5.3

[1] CFP, Chicory flour powder; [2] LA, Lactic acid; [3] SA, Succinic acid; [4] FC, Fold change; [5] Previously described process with BL24 performed under the same conditions by Tsigoriyna et al. [36] was used for comparison. ND, not determined.

For a reliable assessment of the qualities of the mutant BLΔ*sacB* as a producer, a process with 200 g/L soluble inulin at pH 6.25 was carried out. The obtained results and fermentation kinetics were compared with those of the wild-type BL24 as reported by Tsigoriyna et al. [36]. BLΔ*sacB* formed a higher amount of 2,3-BD compared to the wild type (75.5 g/L vs. 67.5 g/L), accompanied by negligible EPS formation. Byproducts remained within close limits, with slightly more glycerol produced (14.1 g/L by BLΔsacB vs. 9.8 g/L by BL24).

Figure 3a shows the kinetics of the process conducted at pH 6.50 with the mutant strain. Converting 300 g/L soluble chicory flour, BLΔ*sacB* formed 128.7 g/L 2,3-BD and 1.8 g/L acetoin. The main byproduct was glycerol (30.2 g/L), followed by a small amount of succinic acid (3.3 g/L). RT-qPCR analysis showed that the genes involved in the hydrolysis of inulin, *sacA*, *sacC*, *fruA*, and *levB*, significantly increased their expression levels, reaching peak values at an mRNA level on the 24th hour (after the exponential phase), then dropping to baseline expression at 48 h (Figure 3b).

At pH 6.75, BLΔ*sacB* reached 96.4 g/L 2,3-BD and 2.1 g/L acetoin, which was some decrease in the amount of target product compared to the process at pH 6.50. This was attributed to a high concentration of lactic acid that was formed in this process (44.8 g/L), and thus, LA became the main byproduct, significantly ahead of glycerol (10.9 g/L). Since at higher pH, the metabolic flux was directed towards LA formation, this result may suggest that the pH optimum for 2,3-BD synthesis was at pH 6.50. For comparison, at pH 6.50, the lactic acid was a temporary product only (Figure 3a). Also, in the same process (pH 6.50), LA reached its maximum of 13.1 g/L at the 24th hour and thereafter was completely assimilated to the 70th hour of the fermentation. At pH 6.75, the visual viscosity of the culture slightly increased notwithstanding the lack of *sacB* expression, thus suggesting that BLΔ*sacB* continued to form EPS different from levan using the glucose and fructose derived from inulin hydrolysis.

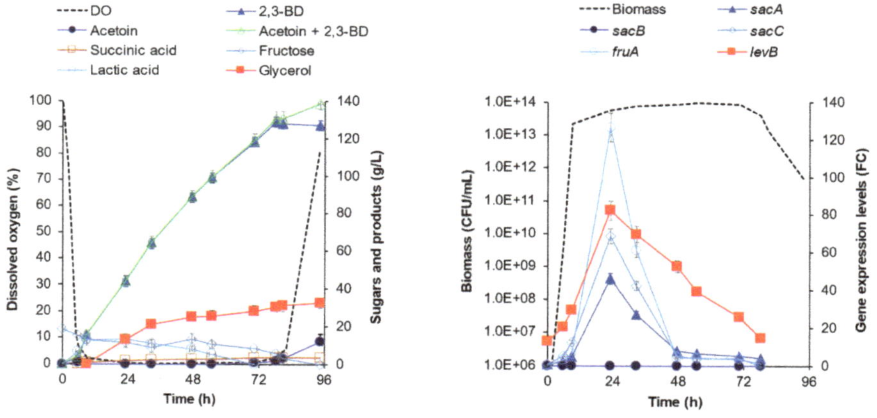

(a) (b)

Figure 3. Production of 2,3-butanediol by BLΔ*sacB* from 300 g/L soluble chicory flour Frutafit® CLR at pH 6.50. (**a**) Time profile of 2,3-BD and byproducts formation, fructose and dissolved oxygen (DO); (**b**) Gene overexpression presented as a fold-change compared to the rate observed at 0 h of the process carried out at pH 5.25 [36]. The fermentation was performed in a bioreactor at 37 °C, with agitation at 500 rpm and aeration at 3.68 vvm.

2.4. Reverse Transcription Real-Time PCR (RT-qPCR)

During fermentation, four genes involved in inulin hydrolysis significantly changed their expression levels at different pH: *sacA*, *sacC*, *fruA*, and *levB* (Figure 4).

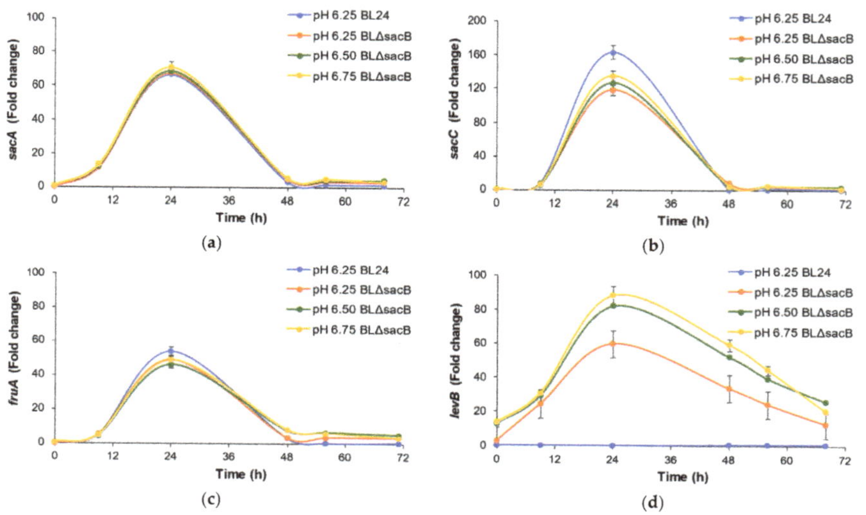

Figure 4. Fold change (FC) in gene expression of *sacA*, *sacC*, *fruA*, and *levB* genes involved in inulin hydrolysis during processes of Frutafit® CLR conversion to 2,3-butanediol by engineered BLΔ*sacB* with pH maintained at the values indicated. FC was calculated vs. 0 h at pH 5.25 of BL24 [36]. (**a**) *sacA*; (**b**) *sacC*; (**c**) *fruA*; (**d**) *levB*.

The lack of expression of *sacB* proves its successful disruption. Compared to the wild-type BL24, *sacA*, *sacC*, and *fruA* attained their upregulated expression at pH 6.25 vs. pH 5.25, conforming to the results shown previously by Tsigoriyna et al. [36]. However, the behavior of *levB* was completely different (Figure 4d). Its expression increased between 60- and 88-fold, depending on the pH of the processes with BLΔ*sacB*, while it remained unaffected in the wild-type BL24. At the 48th h, all gene overexpression (except that of *levB*) entirely declined. The upregulation of *levB* remained high through to the end of the processes, with over 20-fold overexpression in BLΔ*sacB* until the 72nd h.

3. Discussion

A decade ago, a pilot microbial plant in the UK for low-impact production of 2,3-BD with a capacity of 30,000 tons per year was reported to reduce CO_2 emissions by more than 50% [39], thus revealing the significant effect of substituting fermentation for chemical synthesis. The most widely used substrates in industrial biotechnologies for 2,3-BD synthesis are glucose, starch, molasses, and glycerol. According to a techno-economic evaluation performed by Koutinas et al. [11], glycerol is the most profitable, followed by molasses. Since the cost of the raw materials that serve as a carbon source accounts for more than 30% of the total production costs of fermentation, low-price substrates and efficient strains are required for 2,3-BD production [32]. Inulin is identified as an inexpensive, nonfood, renewable source, as it is the third most abundant reserve polysaccharide on the planet (after cellulose and starch). Currently, the price of inulin has leveled off with that of molasses, ranging between 140 and 190 USD per ton [40]. Due to its increasing importance in microbial production on an industrial scale, the global inulin market is expected to reach 1.86 billion USD by the end of 2024. For the period from 2024 to 2029, the demand for inulin is expected to grow at a Compound Annual Growth Rate (CAGR) of 6.55%, with the fastest rate in the Asia-Pacific region and the largest market in Europe, with the global inulin market ultimately reaching 2.56 billion USD by 2029 [41]. On the other hand, the plants from which inulin can be extracted have good productivity (15–50 tons per hectare), and resistance to low temperatures and disease; the time to first grazing is typically 55–85 days after planting [19].

However, a hitherto unsolved problem of industrial fermentations involving *B. licheniformis*, regardless of the target product, is the thickening of the culture due to simultaneous EPS synthesis and uncontrollable foaming during the process [42]. The reason is that the natural strains of *B. licheniformis* simultaneously produce several exopolymers (polyglutamic acid, levan, and heteropolysaccharides) [43], polyols (glycerol and 2,3-BD), and biosurfactants [44]. Levan is the most typical EPS produced by *B. licheniformis* in quantities exceeding 50 g/L (Table 2). Accumulated in high amounts, it increases the viscosity of the fermentation broth and alters some of the process parameters, like aeration and agitation.

Table 2. EPS produced by *B. licheniformis* strains.

Strain	Substrate	EPS	Description	Reference
B. licheniformis	Sucrose	71 g/L	Levan, Δ*epsAB*	[44]
B. licheniformis NS032	Sucrose	53.20 g/L	Levan, optimum pH 7.2	[45]
B. licheniformis 8-37-0-1	Sucrose	47.45 g/L	Levan, optimum pH 6.5–7.0	[46]
B. licheniformis ANT 179	Sugarcane juice	50.25 g/L	Levan, optimum pH 7.0	[47]
BL24	Glucose (fed-batch)	12.61 g/L	EPS of galactose, glucose, and mannose in ratio 54/39/7; pH 6.23	[48]
BL24	Fructose (fed-batch)	7.03 g/L	EPS of glucose, mannose, and galactose in ratio 51/30/19; pH 6.23	[48]

EPS production by *B. licheniformis* depends on the substrate and its quantity [49–51], the fermentation mode (batch or fed-batch) [52,53], and the process parameters, with pH being of key importance [54]. The monosaccharides glucose and fructose contribute to the formation of hetero-EPS, most often consisting of glucose, galactose, and mannose [39,52].

High sucrose concentrations, such as in molasses, provoke levan formation by levansucrase [44–47]. However, during flask-batch processes, regardless of the initial amount of the mono-sugar, EPS synthesis is limited due to the mixed-acid type of fermentation, with lactic and succinic acid among the byproducts. In these processes, the pH of the medium has been observed to spontaneously drop to below pH 5.25, which does not aid the elevated formation of EPS. On the other hand, when the pH maintains a relatively high level (pH 6.0–pH 7.0), it accelerates the formation of both hetero-EPS and levan.

Until now, these phenomena have not been considered in the process of inulin conversion to 2,3-BD by *B. licheniformis*. The Frutafit® CLR chicory flour used in our study contains 80.6% inulin and 19.4% sugars obtained during chicory root processing, of which 7.9% is fructose, 1.5% glucose, and 10% sucrose. Moreover, inulin chains in this substrate have an average degree of polymerization (DP) between 7 and 9, as each of them ends with a sucrose residue, which is an additional substrate for the levansucrase. During the fermentation by the natural strain BL24 at pH 6.25, 197-fold overexpression of *sacB* supported the accumulation of levan in large amounts; therefore, a key strategy to solve this issue is knocking out the *sacB* gene in the wild-type producer.

Some authors focus on the potential benefits of the simultaneous synthesis of levan and 2,3-BD in a mixture [44], but others highlight the difficulties inherent to the subsequent purification steps of either target product and develop engineered strains for less EPS and reduced foaming [55,56]. Mining in the *B. licheniformis* genome revealed that hetero-EPS are produced by a cascade pathway encoded by the operon *epsA-O*, which comprises 16 *eps* genes [57,58]. To reduce EPS formation for efficient alkaline phosphatase production, Zhou et al. [56] knocked out the *eps* cluster in *B. licheniformis* 2709 and disrupted the *lchA-C* genes to diminish the foaming. Applying the same approach, Song et al. [44] accomplished *epsAB* knockout and obtained 88 g/L 2,3-BD together with 70 g/L levan from sucrose.

In our research, the strategy to reduce the total amount of EPS was to inactivate the *sacB* gene encoding levansucrase. This was accomplished by the construction of BLΔ*sacB*, an engineered variant of BL24 containing a disrupted *sacB* gene. The batch fermentation of mono sugars and sucrose showed that *sacB* inactivation resulted in a large reduction in EPS formation from sucrose (more than 15-fold) and a 2.5 times reduction in EPS from glucose and fructose by the recombinant clone compared to the wild type. The overexpression of other genes involved in inulin hydrolysis, *sacA*, *sacC*, and *fruA* encoding sucrases (invertases, EC 3.2.1.26), remained as high as that of the wild type. The *levB* gene encoding levanase (EC 3.2.1.65) showed gradual upregulation in BLΔ*sacB* in a pH-dependent manner and reached its maximum overexpression at pH 6.75. The ~88-fold difference in expression compared to the wild type is possibly due to the changed genomic context after the vector's integration.

Recently, considerable progress has been made in attempts to engage GRAS (Generally Regarded as Safe) bacteria and to utilize different types of inulin for 2,3-BD synthesis [11,30–32]. From acid- or enzyme-hydrolyzed JAT extract by *Paenibacillus polymyxa* ATCC 12321 were obtained 44.0 g/L 2,3-BD [30]; by *P. polymyxa* ZJ-9—37.1 g/L [31]; by *Bacillus* sp. BRC1—28.6 g/L [35], and by *K. pneumoniae* CICC 10011—84.0 g/L [35]. Notably, the highest titer of 2,3-BD was achieved by the type of strain *B. licheniformis* ATCC 14580^T from chicory inulin. Li et al. (2014) performed one-pot saccharification and fermentation with hydrolysis supported by the external additions of recombinant inulinase (SacC), whichyielded 103.0 g/L 2,3-BD in 30 h with a productivity of 3.4 g/L h [32].

In this study, an engineered variant of BL24, distinguished by its ability to convert directly soluble inulin (without the need for preliminary hydrolysis), was developed via *sacB* knockout. The genetically improved BLΔ*sacB* achieved a particularly high titer of 128.7 g/L 2,3-BD. The mutant converted 300 g/L soluble chicory flour in an SSF process and produced the target metabolite with a yield of 0.429 g/g, which is 85.8% of the theoretical yield. The approach of EPS total decrease by *sacB* gene knockout was applied for the first time and apparently is highly effective. By solving the problem of undesired levan accumulation, this study opens the prospect of industrial production of 2,3-BD from inulin-rich substrates by the use of an engineered GRAS producer.

4. Materials and Methods

4.1. Bacterial Strains, Media, and Cultivation Conditions

The wild-type *B. licheniformis* strain 24 is stored in the collection of the Institute of Microbiology, Bulgarian Academy of Sciences, and was identified previously by 16S rRNA gene sequencing (GenBank accession no. MK461938.1) [59].

E. coli HST08 (STELLAR™) competent cells were used in vector constructions and gene cloning (Table 3). The strain was cultivated in LB medium, broth, or supplemented with 15 g/L agar (Alfa Aesar GmbH & Co. KG; Karlsruhe, Germany), and kanamycin with final concentrations of 50 μg/mL (AppliChem GmbH, Darmstadt, Germany).

Table 3. Bacterial strains and plasmids were used in this study.

Strain/Plasmid	Description and Use	Source or Reference
BL24	Natural isolate from a soil sample taken near Yantra River's bed near Veliko Tarnovo, Bulgaria (43°04′52.46″ N 25°37′44.54″ E). Used as a host for *sacB* gene disruption.	[18,36]
E. coli STELLAR™	*F-, endA1, supE44, thi-1, recA1, relA1, gyrA96, phoA, Φ80d lacZΔ M15, Δ(lacZYA-argF) U169, Δ(mrr-hsdRMS-mcrBC), ΔmcrA, λ-*. Used as a host in cloning procedures.	Takara Bio Company (Mountain View, CA, USA)
pBacTag-DYKDDDDK	*B. subtilis* chromosomal integration vector; EryR, AmpR, Epitope FLAG tag. Used for Δ*sacB* disruption cassette construction.	MoBiTec GmbH, Goettingen, Germany
pCR®2.1-TOPO®	*E. coli* TOPO-TA cloning vector. Used as a source of the KanR (NeoR) gene.	Thermo Fisher Scientific Inc., Waltham, MA, USA
BLΔ*sacB*	A mutant of BL24 containing *sacB* gene knockout. Used for 2,3-BD production.	This study
pBac_Kan	Chromosomal integration vector; KanR, AmpR. Used for cloning of Δ*sacB* PCR fragment.	This study
pBac_Kan_Δ*sacB*	Δ*sacB*—containing integrative construct. Used for Δ*sacB* knockout in BL24 chromosomes.	This study

BL24 was maintained on slant LB-agar tubes at 4 °C or as a frozen liquid culture supplemented with 20% glycerol at −70 °C.

Fermentations for 2,3-BD production and EPS synthesis estimation were carried out in the FM (fermentation medium) optimized by Tsigoriyna et al. [18], containing (g/L): sugars (glucose, fructose, sucrose), 50; or CFP (chicory flour powder), 200 or 300; tryptone, 6.41; yeast extract, 13.38; ammonium acetate, 2.5; $(NH_4)_2SO_4$, 1; KH_2PO_4, 3.5; K_2HPO_4, 4.2; $MgSO_4$, 0.32; $CoCl_2{\cdot}6H_2O$, 0.09; microelements solution, 3 mL per liter, containing (g/L): $FeSO_4$, 0.4, H_3BO_3, 0.8; $CuSO_4{\cdot}5H_2O$, 0.04; $NaMoO_4{\cdot}2H_2O$, 0.04; $MnCl_2{\cdot}4H_2O$, 5.0; $ZnSO_4{\cdot}7H_2O$, 0.1; $Co(NO_3)_2{\cdot}6H_2O$, 0.08; $CaCl_2{\cdot}2H_2O$, 1.0; Biotin, 0.01. All chemicals were purchased from Sigma-Aldrich Co. (Saint Louis, MO, USA).

Soluble chicory flour Frutafit® CLR (Sensus B.V., Roosendaal, The Netherlands) was used as a CFP substrate.

Flask-baches were performed as BL24 was cultured in 500 mL Erlenmeyer flasks containing 100 mL of fermentation medium, as the inoculum was a 1% overnight culture. Flask cultivation was on a rotary shaker at 37 °C and 200 rpm.

Batch processes with pH and aeration control were carried out in a 1 L Biostat® A Plus stirred bioreactor (Sartorius Stedim Biotech, Gottingen, Germany), with an additional air pump, rotameter, and vessel for possible excess foam removal. The FM medium described above was used. The process parameters were maintained at optimal values for glucose conversion: temperature 37.8 °C and aeration 3.68 vvm [18]. The pH was maintained at different constant values by adding 6M NaOH or 5M HCl. The added inoculum was 10% (overnight culture of BL24 with OD_{600} = 2.0).

4.2. DNA and RNA Isolation, PCR, and Gibson Assembly Cloning

Total DNA and RNA from samples taken at different hours were isolated with a GeneMATRIX Bacterial & Yeast Genomic DNA Purification Kit and a GeneMATRIX Universal Purification Kit, respectively, according to the instructions of the manufacturer (EURx, Gdansk, Poland).

PCR for fragments shorter than 4 kb was performed with TaKaRa Taq Version 2.0 (Clontech Laboratories, Inc., A Takara Bio Company, Mountain View, CA, USA) in 25 µL reaction volume with 50 ng DNA template and 0.4 µM primers (Table 4). Initial denaturation was set for 1 min at 94 °C, while the hot start was ensured by 10 s at 98 °C each cycle. Elongation time was 1 min per kilobase for each cycle and 5 min final after 35 cycles of amplification. Fragments longer than 4 kb were amplified with TaKaRa LA Taq® DNA Polymerase (Clontech Laboratories, Inc., A Takara Bio Company, Mountain View, CA, USA) in the same reaction volume and with the same amount of template. The reaction mixture also contained freshly added 2.5 mM $MgCl_2$, 0.4 mM from each of the four dNTPs, 0.5 µM of each primer, and 1U of the polymerase. Elongation time was 52 s per kilobase for each cycle and 10 min final after 35 cycles of amplification. All PCR reactions were performed in MultiGene OptiMax Thermal Cycler (Labnet International, Inc., Edison, NJ, USA).

Table 4. Nucleotide sequences of the primers used for gene amplification and/or sequencing. Underlined bases = inserted RBS (Ribosome-binding site). Bases in bold = inserted stop codons. Bases in italics = tails for cloning with Gibson Assembly®.

Primer	Sequence (5′–3′)	PCR Product	Tm (°C)	Molecule Size (bp)
Bac_F	attctatgagtcgcttttgtaaatt	pBacTag_Δ*EryR*	63.8	4738
Bac_R	tgtaatcactccttcttaattacaa		62.4	
Kan_F	*gaaggagtgattaca*aaagagaaagcaggtagcttgc	KanR	65.2	1009
Kan_R	*agcgactcatagaat*tcagaagaactcgtcaagaaggcg		68.5	
BK_F	ggattataaagatgatgatgataaa	pBacTag_KanR	59.2	5747
BK_R	ggtaccctcgactctagat		63.4	
sac_F	*gatctagagtcgagggtacc*tac**taatag**caaggagaagactccctattc	Δ*sacB*	61.0	685
sac_R	*ctttataatccggcc*gaaaattccccgctttattctaag		62.0	

PCR products were visualized on agarose gel (1%) with the addition of fluorescent dye SimplySafe (EURx, Gdansk, Poland) diluted 1:20,000. Sequencing was performed courtesy of Macrogen Inc. (Amsterdam, The Netherlands).

An NEBuilder® HiFi DNA Assembly Cloning Kit (New England Biolabs, Ipswich, MA, USA), an advanced Gibson Assembly cloning system, was used to obtain the recombinant constructs. Ligation reactions were pipetted on ice and consisted of 10 µL NEBuilder HiFi DNA Assembly Master Mix, 200 ng vector, and an amount insert calculated to be in 1:1 molar ratio, and sterile H_2O to 20 µL. Incubation was conducted at 50 °C for 1 h. One-fourth of the reaction was used for the transformation of the NEB 5-alpha competent cells provided with the kit. The mixture of 50 µL competent cells and 5 µL ligation reaction was incubated for 30 min on ice and heat shocked for 50 s at 42 °C. After a brief chill on ice, 950 µL warmed SOC (Super Optimal broth with Catabolite repression; LB with 2% tryptone, 10 mM NaCl, 2.5 mM KCl, 10 mM $MgCl_2$ and 20 mM glucose) was added and the cells were incubated for 1 h at 37 °C and 160 rpm. Transformants were selected after plating 100 µL of the cells on LB-Agar dishes with 50 µg/mL kanamycin and overnight incubation at 37 °C.

4.3. Transformation of *E. coli* and BL24

Transformation of *E. coli* STELLAR^{TM} competent cells (Clontech Laboratories, Inc., A Takara Bio Company, Mountain View, CA, USA) was carried out following Protocol PT5055-2 from the manufacturer. Briefly, 50 µL cells were thawed on ice, mixed with the construct to be transformed, incubated on ice for 30 min, heat shocked for 45 s at 42 °C, and returned to resto on ice for 1–2 min. Warmed SOC medium was added to the final volume of 500 µL and the cells were incubated for 1h at 37 °C and 160 rpm. Cells were then plated on standard Petri dishes with 50 µg/mL kanamycin for the selection of transformants.

BL24 was transformed with a slightly modified version of the high-osmolarity electroporation protocol by Xue et al. [59]. An overnight culture in standard LB medium (1% tryptone, 0.5% yeast extract, 0.5% NaCl), diluted 16 times in LB medium with 0.5 M sorbitol in a 500 mL Erlenmeyer flask, was grown until it reached an OD_{600} of 0.9. The cells were chilled on ice for 10 min and then washed four times with ice-cold electroporation medium (0.5 M sorbitol, 0.5 M mannitol, 10% glycerol). After the last centrifugation (3350 g/10 min/4 °C), the competent cells were resuspended in a final volume of 625 µL electroporation medium, separated into aliquots of 60 µL, and stored at −70 °C. Electroporation was performed in cuvettes with a 0.1 cm electrode gap on a MicroPulser electroporator (BioRad Laboratories, Hercules, CA, USA) with a pulse of 2.1 kV applied for the optimal time of 4–5 ms. Recovery medium (1 mL, 0.5 M sorbitol, and 0.38 M mannitol in LB) was added immediately afterward. The culture was transferred into 15 mL glass tubes and incubated for 3 h at 37 °C. Selection of transformants was obtained on LB-agar plates with 50 µg/mL kanamycin after overnight incubation at 37 °C. The stability of the mutant strain was confirmed by plating it for five passages without antibiotic selective pressure.

4.4. RT-qPCR

Reverse transcription (RT) was accomplished with the NG dART RT Mix (EURx, Gdansk, Poland). The 20 µL reactions contained 1 µg total RNA and 200 ng random hexamer primers and were subjected to the following program: 10 min at 25 °C for primer hybridization, 50 min at 50 °C for the reverse transcription itself, and 5 min at 85 °C for the inactivation of the enzyme. Before the reverse transcription, all RNA samples were subjected to DNase I (5U) treatment in a buffer with 25 mM $MgCl_2$ (EURx, Gdansk, Poland) for 30 min at 37 °C. Enzyme inactivation was achieved for 10 min at 65 °C with the addition of 20 mM EDTA. RT-qPCR was performed with SsoFast^{TM} EvaGreen® Supermix with Low ROX (Bio-Rad, Hercules, CA, USA) in a Corbett Research RG-6000 Real-Time PCR Thermocycler (Qiagen, Germantown, MD, USA). Primer pairs were designed to amplify fragments from 62 to 125 bp within *sacA*, *sacB*, *sacC*, *levB*, *fruA*, and the control 16S rRNA gene (Table 5).

Table 5. Primers used in RT-qPCR experiments.

Primer	Sequence (5′–3′)	PCR Product (bp)	Position in Gene *
16S_F 16S_R	gagtacgaccgcaaggttga cctggtaaggttcttcgcgt	100	875–895 975–955
sacA_RTF sacA_RTR	aagagatcgccctcacgccgagcgactggttt atttccctcgccgtctctgacattccccgtgt	125	255–286 379–348
sacB_RTF sacB_RTR	caacagagcctactacgggggcagcaagaagt tcgatgattccgagagcgccgttagccagcga	117	861–892 977–946
sacC_RTF sacC_RTR	gccgctcgttgccatttatacgcaggaccgga gctgtaggcgatgctttgcacttgttccccgc	64	375–406 438–407
levB_RTF levB_RTR	gcatactggacaggcagcttcaacggcaacga cgttcgtttcgccgtcctcaaatgtcacgccc	121	784–815 904–873
fruA_RTF fruA_RTR	gggagtcagagatgccgacgaaagcagacgga ttcacgcggcaaagttaatgccccgcaccatc	62	893–924 954–923

* Positions are according to the respective gene sequences *B. licheniformis* ATCC 14580^{T} (NCBI GenBank acc. no. CP034569.1).

The optimal annealing temperature was determined to be 60 °C. Each reaction of 20 μL contained 40 ng cDNA as a template and 500 nM primers. 16S rRNA was used as an internal control for each sample in each run. The beginning (0 h) at the lowest pH (5.2) was used as a basis for comparison. Relative expression was calculated by the ΔΔCt method as described previously [36]. The presented results are the mean values of three independent fermentation experiments, each tested by three independent RT-PCR trials. The standard deviation was 6%.

4.5. Analytical Methods

The cell growth of the inoculum cultures was monitored by measuring the optical density (OD) at wavelength 600 using a UV/VIS Spectrophotometer (Thermo Scientific Inc., Waltham, MA, USA).

The viable cell counts (CFU, colony-forming units, per mL) were estimated as decimal dilutions of samples plated on LB-agar plates.

RNA concentrations and purity (Abs_{260}/Abs_{280} ratio) were determined using Quawell UV Spectrophotometer Q3000 (Quawell Technology, Inc., San Jose, CA, USA).

The concentrations of the obtained 2,3-BD, acetoin, glycerol, lactic, and succinic acids were analyzed with a YL Instrument 9300 HPLC System (YL Instrument Co., Ltd., Anyang, Republic of Korea) equipped with an RI detector (YL 9170), using HPLC column Aminex HPX-87H at 65 °C with a mobile phase of 5 mM of H_2SO_4 at a flow rate of 0.6 mL/min (BioRad Laboratories, Hercules, CA, USA). For glucose, fructose, and sucrose quantification, a column HPX-87C (BioRad Laboratories, Hercules, CA, USA) at 85 °C was used. As a mobile phase, water with a flow rate of 0.6 mL/min was used.

All EPSs formed from glucose, fructose, and sucrose were extracted by the following procedure. Crude EPS fractions were isolated from the fermentation medium after initial centrifugation at 6000× *g* for 30 min to remove the biomass. The supernatant was deproteinated by incubation with 14% trichloroacetic acid (Merck KGaA, Darmstadt, Germany) in a rotary shaker (90 rpm) at 37 °C for 40 min. The sample was then centrifuged at 10,000× *g* for 10 min, at 4 °C, to remove the denatured proteins. The supernatant (crude EPS) was precipitated against three volumes of ice-cold ethanol (96%) and incubated at −18 °C overnight. The EPSs were harvested by centrifugation at 10,000× *g* for 20 min, washed twice with 50% ethanol, and the pellet was air-dried and dissolved in sterile dd H_2O. After overnight dialysis at 4 °C, the sample was dried in a desiccator and stored at 4 °C. Bradford assay showed that all EPSs did not contain any amounts of residual protein. The carbohydrate content in each EPS was tested with the phenol-sulfuric acid colorimetric method of Dubois [60].

5. Conclusions

The present study reports a record amount of 2,3-BD produced from inulin involving the engineered BLΔ*sacB*, a variant of the natural isolate BL24. By disrupting a single gene, *sacB*, encoding levansucrase, the mutant BLΔ*sacB* produced 15-fold less levan from sucrose than the wild type. Thus, when the fermentation was carried out at the optimum process pH of 6.50, 300 g/L soluble chicory flour was directly converted to 128.7 g/L 2,3-BD, with a productivity of 1.65 g/L/h and a yield of 0.43 g/g, which is close to the theoretical maximum. The two-fold increase in 2,3-BD titer obtained from the engineered strain compared to the wild type reveals the great potential of the mutation in the *sacB* gene to solve the problem of undesired levan accumulation by *B. licheniformis*.

Author Contributions: Conceptualization, K.P.; methodology, P.P., K.P. and A.A.; software, E.G.; investigation, L.T., I.I., N.A., E.G. and A.A.; resources, K.P.; writing—original draft preparation, P.P. and K.P.; writing—review and editing, K.P.; funding acquisition, P.P. and K.P. All authors have read and agreed to the published version of the manuscript.

Funding: The authors kindly acknowledge the financial support of project № BG05M2OP001-1.002-0014 "Center of competence HITMOBIL–Technologies and systems for generation, storage, and consumption of clean energy", funded by Operational Programme "Science and Education for Smart Growth" 2014–2020, co-funded by the EU from European Regional Development Fund.

Institutional Review Board Statement: Not applicable.

Informed Consent Statement: Not applicable.

Data Availability Statement: Nucleotide sequences are available in the NCBI GenBank. The strain BL24 and its engineered variant can be obtained from the authors on request.

Conflicts of Interest: The authors declare no conflicts of interest.

References

1. Muras, A.; Romero, M.; Mayer, C.; Otero, A. Biotechnological applications of *Bacillus licheniformis*. *Crit. Rev. Biotechnol.* **2021**, *41*, 609–627. [CrossRef] [PubMed]
2. Kumar, P.; Patel, S.K.; Lee, J.K.; Kalia, V.C. Extending the limits of *Bacillus* for novel biotechnological applications. *Biotechnol. Adv.* **2013**, *31*, 1543–1561. [CrossRef] [PubMed]
3. Martín-González, D.; Bordel, S.; Solis, S.; Gutierrez-Merino, J.; Santos-Beneit, F. Characterization of *Bacillus* Strains from Natural Honeybee Products with High Keratinolytic Activity and Antimicrobial Potential. *Microorganisms* **2023**, *11*, 456. [CrossRef] [PubMed]
4. Shleeva, M.O.; Kondratieva, D.A.; Kaprelyants, A.S. *Bacillus licheniformis*: A Producer of Antimicrobial Substances, including Antimycobacterials, Which Are Feasible for Medical Applications. *Pharmaceutics* **2023**, *15*, 1893. [CrossRef]
5. Maina, S.; Prabhu, A.A.; Vivek, N.; Vlysidis, A.; Koutinas, A.; Kumar, V. Prospects on bio-based 2,3-butanediol and acetoin production: Recent progress and advances. *Biotechnol. Adv.* **2022**, *54*, 107783. [CrossRef] [PubMed]
6. Song, D.; Cho, S.-Y.; Vu, T.-T.; Duong, H.-P.-Y.; Kim, E. Dehydration of 2,3-Butanediol to 1,3-Butadiene and Methyl Ethyl Ketone: Modeling, Numerical Analysis and Validation Using Pilot-Scale Reactor Data. *Catalysts* **2021**, *11*, 999. [CrossRef]
7. Li, L.X.; Li, K.; Wang, K.; Chen, C.; Gao, C.; Ma, C.Q.; Xu, P. Efficient production of 2,3-butanediol from corn stover hydrolysate by using a thermophilic *Bacillus licheniformis* strain. *Bioresour. Technol.* **2014**, *170*, 256–261. [CrossRef]
8. Lü, C.; Ge, Y.; Cao, M.; Guo, X.; Liu, P.; Gao, C.; Xu, P.; Ma, C. Metabolic Engineering of *Bacillus licheniformis* for Production of Acetoin. *Front. Bioeng. Biotechnol.* **2020**, *8*, 125. [CrossRef]
9. Tinôco, D.; Seldin, L.; Coutinho, P.; Freire, D.M.G. Sustainable production of optically pure platform chemical bio-based (R,R)−2,3-butanediol from sugarcane molasses in a low-cost salt medium. *Ind. Crops Prod.* **2024**, *209*, 117931. [CrossRef]
10. Rebecchi, S.; Pinelli, D.; Zanaroli, G.; Fava, F.; Frascari, D. Effect of oxygen mass transfer rate on the production of 2,3-butanediol from glucose and agro-industrial byproducts by *Bacillus licheniformis* ATCC 9789. *Biotechnol. Biofuels* **2018**, *11*, 145. [CrossRef]
11. Koutinas, A.A.; Yepez, B.; Kopsahelis, N.; Freire, D.M.G.; de Castro, A.M.; Papanikolaou, S.; Kookos, I.K. Techno-economic evaluation of a complete bioprocess for 2,3-butanediol production from renewable resources. *Bioresour. Technol.* **2016**, *204*, 55–64. [CrossRef]
12. Dai, J.-Y.; Guan, W.-T.; Xiu, Z.-L. Bioconversion of inulin to 2,3-butanediol by a newly isolated *Klebsiella pneumoniae* producing inulinase. *Process Biochem.* **2020**, *98*, 247–253. [CrossRef]
13. Petrov, K.; Petrova, P. Current Advances in Microbial Production of Acetoin and 2,3-Butanediol by *Bacillus* spp. *Fermentation* **2021**, *7*, 307. [CrossRef]
14. Sun, R.; Kang, J.; Wang, X.; Xiu, B.; Ping, W.; Ge, J. Enhancement of 2,3-Butanediol Production by *Klebsiella pneumoniae*: Emphasis on the Mediation of sRNA-SgrS on the Carbohydrate Utilization. *Fermentation* **2022**, *8*, 359. [CrossRef]
15. Palaiogeorgou, A.M.; Delopoulos, E.I.M.; Koutinas, A.A.; Papanikolaou, S. Screening Bacterial Strains Capable of Producing 2,3-Butanediol: Process Optimization and High Diol Production by *Klebsiella oxytoca* FMCC-197. *Fermentation* **2023**, *9*, 1014. [CrossRef]
16. Ju, J.-H.; Jo, M.-H.; Heo, S.-Y.; Kim, M.-S.; Kim, C.-H.; Paul, N.C.; Sang, H.; Oh, B.-R. Production of highly pure R,R-2,3-butanediol for biological plant growth promoting agent using carbon feeding control of *Paenibacillus polymyxa* MDBDO. *Microb. Cell Factories* **2023**, *22*, 121. [CrossRef] [PubMed]
17. Jurchescu, I.M.; Hamann, J.; Zhou, X.; Ortmann, T.; Kuenz, A.; Prusse, U.; Lang, S. Enhanced 2,3-butanediol production in fed batch cultures of free and immobilized *Bacillus licheniformis* DSM 8785. *Appl. Microbiol. Biotechnol.* **2013**, *97*, 6715–6723. [CrossRef] [PubMed]
18. Tsigoriyna, L.; Ganchev, D.; Petrova, P.; Petrov, K. Highly Efficient 2,3-Butanediol Production by *Bacillus licheniformis* via Complex Optimization of Nutritional and Technological Parameters. *Fermentation* **2021**, *7*, 118. [CrossRef]
19. Guo, Z.-W.; Ni, Z.-F.; Zong, M.-H.; Lou, W.-Y. Modular Metabolic Engineering of *Bacillus licheniformis* for Efficient 2,3-Butanediol Production by Consolidated Bioprocessing of Jerusalem Artichoke Tubers. *ACS Sustain. Chem. Eng.* **2022**, *10*, 9624–9634. [CrossRef]
20. Petrova, M.; Miladinova-Georgieva, K.; Geneva, M. Influence of Abiotic and Biotic Elicitors on Organogenesis, Biomass Accumulation, and Production of Key Secondary Metabolites in Asteraceae Plants. *Int. J. Mol. Sci.* **2024**, *25*, 4197. [CrossRef]
21. Alonso-Allende, J.; Milagro, F.I.; Aranaz, P. Health Effects and Mechanisms of Inulin Action in Human Metabolism. *Nutrients* **2024**, *16*, 2935. [CrossRef] [PubMed]

22. China Chicory Prices. Available online: https://www.selinawamucii.com/insights/prices/china/chicory/ (accessed on 12 October 2024).
23. Singh, R.S.; Singh, T.; Hassan, M.; Kennedy, J.F. Updates on inulinases: Structural aspects and biotechnological applications. *Int. J. Biol. Macromol.* **2020**, *164*, 193–210. [CrossRef] [PubMed]
24. Singh, R.S.; Chauhan, K.; Kennedy, J.F. A panorama of bacterial inulinases: Production, purification, characterization and industrial applications. *Int. J. Biol. Macromol.* **2017**, *96*, 312–322. [CrossRef]
25. Petrov, K.; Popova, L.; Petrova, P. High lactic acid and fructose production via Mn^{2+}-mediated conversion of inulin by *Lactobacillus paracasei*. *Appl. Microbiol. Biotechnol.* **2017**, *101*, 4433–4445. [CrossRef]
26. Lara-Fiallos, M.; Ayala Chamorro, Y.T.; Espín-Valladares, R.; DelaVega-Quintero, J.C.; Olmedo-Galarza, V.; Nuñez-Pérez, J.; Pais-Chanfrau, J.-M.; Martínez, A.P. Immobilised Inulinase from *Aspergillus niger* for Fructose Syrup Production: An Optimisation Model. *Foods* **2024**, *13*, 1984. [CrossRef]
27. Hughes, S.R.; Qureshi, N.; López-Núñez, J.C.; Jones, M.A.; Jarodsky, J.M.; Galindo-Leva, L.Á.; Lindquist, M.R. Utilization of inulin-containing waste in industrial fermentations to produce biofuels and bio-based chemicals. *World J. Microbiol. Biotechnol.* **2017**, *33*, 78. [CrossRef] [PubMed]
28. Patakova, P.; Linhova, M.; Rychtera, M.; Paulova, L.; Melzoch, K. Novel and neglected issues of acetone-butanol-ethanol (ABE) fermentation by clostridia: Clostridium metabolic diversity, tools for process mapping and continuous fermentation systems. *Biotechnol. Adv.* **2013**, *31*, 58–67. [CrossRef]
29. Sun, L.H.; Wang, X.D.; Dai, J.Y.; Xiu, Z.L. Microbial production of 2,3-butanediol from Jerusalem artichoke tubers by *Klebsiella pneumoniae*. *Appl. Microbiol. Biotechnol.* **2009**, *82*, 847–852. [CrossRef] [PubMed]
30. Fages, J.; Mulard, D.; Rouquet, J.J.; Wilhelm, J.L. 2,3-Butanediol production from Jerusalem artichoke, Helianthus Tuberosus, by *Bacillus polymyxa* ATCC 12321. Optimization of kL a profile. *Appl. Microbiol. Biotechnol.* **1986**, *25*, 197–202. [CrossRef]
31. Gao, J.; Xu, H.; Li, Q.-J.; Feng, X.-H.; Li, S. Optimization of medium for one-step fermentation of inulin extract from Jerusalem artichoke tubers using *Paenibacillus polymyxa* ZJ-9 to produce R, R-2,3-butanediol. *Bioresour. Technol.* **2010**, *101*, 7087–7093. [CrossRef]
32. Li, L.X.; Chen, C.; Li, K.; Wang, Y.; Gao, C.; Ma, C.Q.; Xu, P. Efficient simultaneous saccharification and fermentation of inulin to 2,3-butanediol by thermophilic *Bacillus licheniformis* ATCC 14580. *Appl. Environ. Microbiol.* **2014**, *80*, 6458–6464. [CrossRef] [PubMed]
33. Song, C.W.; Rathnasingh, C.; Park, J.M.; Lee, J.; Song, H. Isolation and evaluation of *Bacillus* strains for industrial production of 2,3-butanediol. *J. Microbiol. Biotechnol.* **2018**, *28*, 409–417. [CrossRef] [PubMed]
34. Tsigoriyna, L.; Arsov, A.; Petrova, P.; Gergov, E.; Petrov, K. Heterologous Expression of Inulinase Gene in *Bacillus licheniformis* 24 for 2,3-Butanediol Production from Inulin. *Catalysts* **2023**, *13*, 841. [CrossRef]
35. Park, J.M.; Oh, B.-R.; Kang, I.Y.; Heo, S.-Y.; Seo, J.-W.; Park, S.-M.; Hong, W.-K.; Kim, C.H. Enhancement of 2,3-butanediol production from Jerusalem artichoke tuber extract by a recombinant *Bacillus* sp. strain BRC1 with increased inulinase activity. *J. Ind. Microbiol. Biotechnol.* **2017**, *44*, 1107–1113. [CrossRef] [PubMed]
36. Tsigoriyna, L.; Arsov, A.; Gergov, E.; Petrova, P.; Petrov, K. Influence of pH on Inulin Conversion to 2,3-Butanediol by *Bacillus licheniformis* 24: A Gene Expression Assay. *Int. J. Mol. Sci.* **2023**, *24*, 14065. [CrossRef]
37. Doan, C.T.; Tran, T.N.; Nguyen, T.T.; Tran, T.P.H.; Nguyen, V.B.; Tran, T.D.; Nguyen, A.D.; Wang, S.L. Production of Sucrolytic Enzyme by *Bacillus licheniformis* by the Bioconversion of Pomelo Albedo as a Carbon Source. *Polymers* **2021**, *13*, 1959. [CrossRef]
38. Domżał-Kędzia, M.; Ostrowska, M.; Lewińska, A.; Łukaszewicz, M. Recent Developments and Applications of Microbial Levan, A Versatile Polysaccharide-Based Biopolymer. *Molecules* **2023**, *28*, 5407. [CrossRef]
39. Novamont: Opening of the World's First Industrial Scale Plant for the Production of Butanediol via Fermentation of Renewable Raw Materials. Available online: https://uk.novamont.com/leggi_press.php?id_press=84 (accessed on 30 October 2024).
40. Inulin Suppliers. Available online: https://www.made-in-china.com/manufacturers/inulin.html (accessed on 30 October 2024).
41. Inulin Market Size & Share Analysis. Available online: https://www.mordorintelligence.com/industry-reports/inulin-market (accessed on 30 October 2024).
42. Leal, E.; Teixeira, J.A.; Gudiña, E.J. Development of Foam-Free Biosurfactant Production Processes Using *Bacillus licheniformis*. *Fermentation* **2024**, *10*, 340. [CrossRef]
43. Wei, X.; Yang, L.; Chen, Z.; Xia, W.; Chen, Y.; Cao, M.; He, N. Molecular weight control of poly-γ-glutamic acid reveals novel insights into extracellular polymeric substance synthesis in *Bacillus licheniformis*. *Biotechnol. Biofuels* **2024**, *17*, 60. [CrossRef]
44. Song, C.W.; Kwon, M.; Song, H. Simultaneous and selective production of exopolymers and polyols by metabolically engineered *Bacillus licheniformis* strains. *Biochem. Eng. J.* **2022**, *181*, 108381. [CrossRef]
45. Gojgic-Cvijovic, G.D.; Jakovljevic, D.M.; Loncarevic, B.D.; Todorovic, N.M.; Pergal, M.V.; Ciric, J.; Loos, K.; Beskoski, V.P.; Vrvic, M.M. Production of levan by *Bacillus licheniformis* NS032 in sugar beet molasses-based medium. *Int. J. Biol. Macromol.* **2019**, *121*, 142–151. [CrossRef] [PubMed]
46. Liu, C.; Lu, J.; Lu, L.; Liu, Y.; Wang, F.; Xiao, M. Isolation, structural characterization and immunological activity of an exopolysaccharide produced by *Bacillus licheniformis* 8-37-0-1. *Bioresour. Technol.* **2010**, *101*, 5528–5533. [CrossRef] [PubMed]
47. Xavier, J.R.; Ramana, K.V. Optimization of Levan Production by Cold-Active *Bacillus licheniformis* ANT 179 and Fructooligosaccharide Synthesis by Its Levansucrase. *Appl. Biochem. Biotechnol.* **2017**, *181*, 986–1006. [CrossRef]

48. Petrova, P.; Arsov, A.; Ivanov, I.; Tsigoriyna, L.; Petrov, K. New Exopolysaccharides Produced by *Bacillus licheniformis* 24 Display Substrate-Dependent Content and Antioxidant Activity. *Microorganisms* **2021**, *9*, 2127. [CrossRef]
49. Öner, E.T.; Hernández, L.; Combie, J. Review of Levan polysaccharide: From a century of past experiences to future prospects. *Biotechnol. Adv.* **2016**, *34*, 827–844. [CrossRef]
50. Vinothkanna, A.; Sathiyanarayanan, G.; Balaji, P.; Mathivanan, K.; Pugazhendhi, A.; Ma, Y.; Sekar, S.; Thirumurugan, R. Structural characterization, functional and biological activities of an exopolysaccharide produced by probiotic *Bacillus licheniformis* AG-06 from Indian polyherbal fermented traditional medicine. *Int. J. Biol. Macromol.* **2021**, *174*, 144–152. [CrossRef]
51. Xu, Z.; Chen, G.; Xue, L.; Zhang, H.; Wang, J.; Xiang, H.; Lia, J.; Zheng, K. Isolation, structural characterizations and bioactivities of exopolysaccharides produced by *Bacillus licheniformis*. *Int. J. Biol. Macromol.* **2019**, *141*, 298–306. [CrossRef]
52. Nguyen, H.-T.; Pham, T.-T.; Nguyen, P.-T.; Le-Buanec, H.; Rabetafika, H.N.; Razafindralambo, H.L. Advances in Microbial Exopolysaccharides: Present and Future Applications. *Biomolecules* **2024**, *14*, 1162. [CrossRef] [PubMed]
53. Białkowska, A.M.; Jedrzejczak-Krzepkowska, M.; Gromek, E.; Krysiak, J.; Sikora, B.; Kalinowska, H.; Kubik, C.; Schütt, F.; Turkiewicz, M. Effects of genetic modifications and fermentation conditions on 2,3-butanediol production by alkaliphilic *Bacillus subtilis*. *Appl. Microbiol. Biotechnol.* **2016**, *100*, 2663–2676. [CrossRef]
54. Porras-Domínguez, J.R.; Ávila-Fernández, Á.; Rodríguez-Alegría, M.E.; Miranda-Molina, A.; Escalante, A.; González-Cervantes, R.; Olvera, C.; López-Munguía, A. Levan-type FOS production using a *Bacillus licheniformis* endolevanase. *Process Biochem.* **2014**, *49*, 783–790. [CrossRef]
55. Fan, Y.; Wang, J.; Gao, C.; Zhang, Y.; Du, W. A novel exopolysaccharide-producing and long-chain n-alkane degrading bacterium *Bacillus licheniformis* strain DM-1 with potential application for in-situ enhanced oil recovery. *Sci. Rep.* **2020**, *10*, 8519. [CrossRef] [PubMed]
56. Zhou, C.; Zhou, H.; Li, D.; Zhang, H.; Wang, H.; Lu, F. Optimized expression and enhanced production of alkaline protease by genetically modified *Bacillus licheniformis* 2709. *Microb. Cell Factories* **2020**, *19*, 45. [CrossRef] [PubMed]
57. Chen, Z.; Liu, P.; Li, Z.; Yu, W.; Wang, Z.; Yao, H.; Wang, Y.; Li, Q.; Deng, X.; He, N. Identification of key genes involved in polysaccharide bioflocculant synthesis in *Bacillus licheniformis*. *Biotechnol. Bioeng.* **2017**, *114*, 645–655. [CrossRef]
58. Rey, M.W.; Ramaiya, P.; Nelson, B.A.; Brody-Karpin, S.D.; Zaretsky, E.; Tang, M.; Lopez de Leon, A.; Xiang, H.; Gusti, V.; Clausen, I.G.; et al. Complete genome sequence of the industrial bacterium *Bacillus licheniformis* and comparisons with closely related *Bacillus* species. *Genome Biol.* **2004**, *5*, R77. [CrossRef]
59. Xue, G.-P.; Johnson, J.S.; Dalrymple, B.P. High osmolarity improves the electro-transformation efficiency of the gram-positive bacteria *Bacillus subtilis* and *Bacillus licheniformis*. *J. Microbiol. Methods* **1999**, *34*, 183–191. [CrossRef]
60. Dubois, M.; Gilles, K.A.; Hamilton, J.K.; Rebers, P.A.; Smith, F. Colorimetric method for determination of sugars and related substances. *Anal. Chem.* **1956**, *28*, 350–356. [CrossRef]

International Journal of
Molecular Sciences

MDPI

Article

ARTP/NTG Compound Mutagenesis Improved the Spinosad Production and the Insecticidal Virulence of *Saccharopolyspora Spinosa*

Zirong Zhu †, Wangqiong Chen †, Li Cao, Ziyuan Xia, Jie Rang, Shengbiao Hu and Liqiu Xia *

State Key Laboratory of Developmental Biology of Freshwater Fish, Hunan Provincial Key Laboratory of Microbial Molecular Biology, College of Life Science, Hunan Normal University, Changsha 410081, China; 201820141158@hunnu.edu.cn (Z.Z.); 202320142772@hunnu.edu.cn (W.C.); 201920142230@hunnu.edu.cn (L.C.); 201910140184@hunnu.edu.cn (Z.X.); rang0214@hunnu.edu.cn (J.R.); shengbiaohu@hunnu.edu.cn (S.H.)

* Correspondence: xialq@hunnu.edu.cn

† These authors contributed equally to this study.

Abstract: Spinosad is an efficient and broad-spectrum environmentally friendly biopesticide, but its low yield in wild-type *Saccharopolyspora spinosa* limits its further application. ARTP/NTG compound mutagenesis was used in this study to improve the spinosad titer of *S. spinosa* and obtain a high-yield mutant—NT24. Compared with the wild-type strain, the fermentation cycle of NT24 was shortened by 2 days and its maximum titer of spinosad reached 858.3 ± 27.7 mg/L, which is 5.12 times more than for the same-period titer of the wild-type strain. In addition, RT-qPCR, resequencing, and targeted metabolomics showed that the upregulation of the key differential genes *accD6*, *fadD*, *sdhB*, *oadA*, and *gntZ* caused increased metabolic flux in the tricarboxylic acid cycle and pentose phosphate pathway, suggesting that the accumulation of pyruvate and short-chain acyl-CoA was the primary cause of spinosad accumulation in NT24. This study demonstrates the effectiveness of ARTP mutagenesis in *S. spinosa*, and provides new insights for the mechanism of spinosad biosynthesis and metabolic engineering in *S. spinosa*.

Keywords: *Saccharopolyspora spinosa*; ARTP/NTG mutagenesis; spinosad; biotechnology

1. Introduction

Saccharopolyspora spinosa (*S. spinosa*), a Gram-positive filamentous bacterium, is a rare actinomycete and the source of a variety of secondary metabolites with bioactivities, such as macrolides [1–3]. *S. spinosa* was originally isolated from soil samples from the Caribbean islands in 1982, and is capable of producing macrolide secondary metabolites with high insecticidal activity after aerobic fermentation [4,5]. These bioactive substances have been defined as spinosad [6]. A macrolide compound, spinosad is composed of a twelve-membered lactone ring, a 5-6-5 cis-trans-cis tricyclic ring, and two glycosyl groups [7]. Due to the different positions of acyl modification on the glycosyl group, spinosad has a variety of homologues in structure [8]. At least twenty-five types of spinosyn were isolated from a fermentation broth of *S. spinosa*, in which the main components were spinosyn A and spinosyn D [9]. The mixture of spinosyn A and spinosyn D was called spinosad [10], which has highly specific and insecticidal activity toward a variety of pests, e.g., Lepidoptera [11], Diptera [12], and Thysanoptera [13], as well as toxicity toward Coleoptera and Hymenoptera [14]. Spinosad mainly acts on the insect nervous system, especially the acetylcholine receptor and γ-aminobutyric acid (GABA) receptor [15]. After binding to these receptors, the nerve conduction of the target pest will be disturbed, thus causing persistent muscle contraction and nerve overexcitation and ultimately leading to paralysis and death of the target pest [16,17]. In addition, compared with traditional pesticides, spinosad leaves little environmental residue [18,19]; therefore, it has been widely used in food storage and crop pest control [20].

However, it is difficult to screen novel *S. spinosa* strains by conventional soil separation methods, and the spinosad production of the wild-type *S. spinosa* strains obtained from the natural environment is often too low to meet the needs of large-scale industrial production. Therefore, various strategies have been adopted to increase the production capacity of wild-type *S. spinosa* to enable it to adapt to industrial production in recent years [21], such as metabolic engineering [22], transcriptomics-based genetic modification [23], medium optimization [4], and random mutation [24]. Although various gene modification methods are the first choice for improving the performance of strains, in many cases, random mutation breeding is still the most effective way to change the genes and characteristics of the microbiome [25].

The *S. spinosa* genome contains high G + C content and has a large number of biosynthetic gene clusters. Using traditional genetic modification on it faces challenges such as difficulty in operation, long cycle, and unpredictable results, and the cost of trial and error is extremely high [26,27]. From a commercial standpoint, there are many limitations to the application of genetically modified products, and the strains obtained from random mutagenesis are always classified as non-genetic modifications [28]. Thus, artificial mutagenesis methods, such as physical mutagenesis, chemical mutagenesis, and physical–chemical compound mutagenesis, are often employed to accelerate the evolution of strains to obtain strains with improved performance. For physical mutagenesis, ARTP (atmospheric and room temperature plasma) mutagenesis is an emerging mutagenesis technology [29]. In the process of ARTP mutagenesis, the excited particles in plasma jets act on the strains and induce damage to the cell wall, cell membrane, and DNA or RNA, leading to changes in cell membrane permeability and the activity of various intracellular proteins, which may trigger the initiation of SOS repair (the source of mutations and new phenotypes) [30–32].

Compared with traditional physical mutagenesis, such as ultraviolet irradiation and gamma rays, ARTP uses a non-invasive method to mutate the strain, which is safer and can cause a higher degree of DNA damage [28,32]. In addition, ARTP mutagenesis is driven by RF power, which can be produced under normal pressure and room temperature without expensive vacuum systems or extreme temperature. The mutation distance between the plasma trigger and samples, as well as the intensity and duration of plasma jets, can be adjusted, rendering it applicable to various strains [33,34]. At present, ARTP mutagenesis has been applied to the breeding of a variety of bacteria [35–37].

In this study, we first employed ARTP/NTG compound mutagenesis to improve the spinosad production of *S. spinosa*. To further reveal the high-yield mechanism of spinosad in *S. spinosa*, transcription-level verification combined with resequencing and target metabolomic analysis was conducted. This study not only aimed to increase the yield of spinosad, but also to provide a presumptive mechanism of spinosad biosynthesis promotion in *S. spinosa*.

2. Results

2.1. ARTP/NTG Compound Mutagenesis Improved the Titer of Spinosad in S. spinosa

To determine the optimal mutation time of ARTP mutagenesis, the lethality rates of wild-type *S. spinosa* (WT) were assessed under exposure times of 0–150 s (0, 25, 50, 75, 90, 100, 125, and 150 s) (Figure 1A). As shown in Figure 1, a dose–response relationship was observed between the lethality rate of WT and ARTP treatment duration. After 90 s and 100 s exposure under ARTP, the lethality rate reached 89% and 93.6%, respectively. Consequently, 179 mutants obtained from 90 s and 100 s ARTP mutagenesis were subjected to batch fermentation for yield determination, with mutant A5 exhibiting the greatest increase in spinosad production, yielding a titer of 597.4 ± 14.5 mg/L, 1.8 times that of the WT strain (317.5 ± 12.7 mg/L) (Figure 1B).

Thus, A5 was selected as the initial strain of NTG mutagenesis. The lethality rates of A5 were determined under different NTG concentrations (0, 100, 200, 300, 400, and 500 μg/mL) to identify the optimal concentration for NTG mutagenesis (Figure 1C). The results show that 300 μg/mL NTG mutagenesis results in a lethality rate of 88.3%. After 1 h of mutagenesis with 300 μg/mL NTG, three mutants (NT24, NT42 and NT60) exhibiting improved spinosad titer were screened. Among them, the spinosad titer of mutant NT24 was the highest (858.3 ± 27.7 mg/L), which is 2.58 and 1.43 times higher than those of the WT strain and mutant A5, respectively (Figure 1B). The fermentation products were collected for mass spectrometry identification, and the mass spectrometry (MS) results revealed the [M + -H] + ions at m/z = 732.4 (Figure S1). Additionally, mutant NT24 was confirmed to exhibit stable spinosad production after five subcultures (Figure 1D).

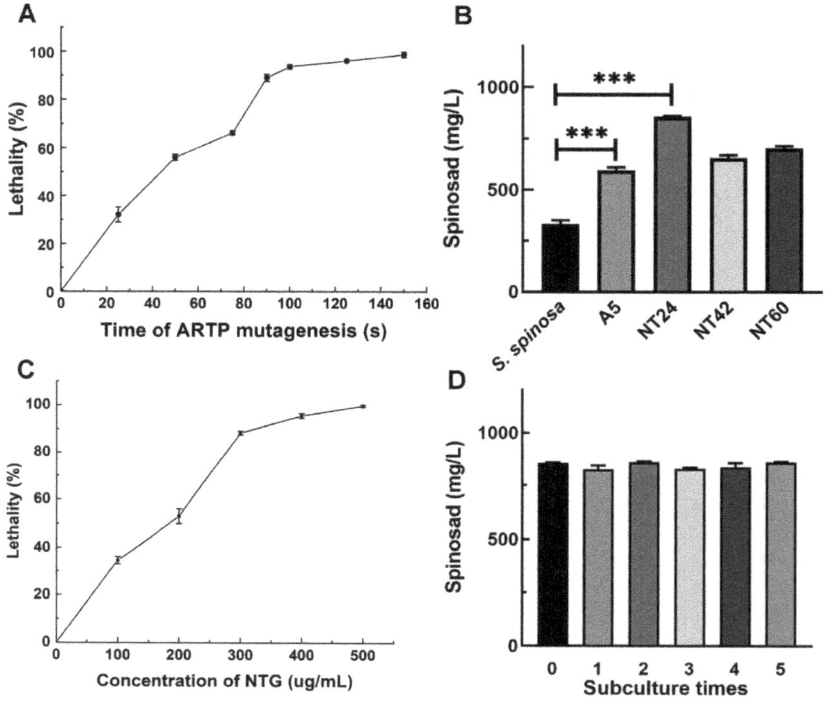

Figure 1. ARTP/NTG compound mutagenesis in *S. spinosa*. (**A**) The lethality rates of *S. spinosa* under different durations of ARTP treatment. (**B**) Spinosad titer of high-yield mutants of *S. spinosa* from ARTP/NTG compound mutagenesis. (**C**) The lethality rates of *S. spinosa* under different concentrations of NTG for 1 h. (**D**) Subculture of NT24 and spinosad titer of each passage. Error bars show standard deviations. Univariate variance, *** $p < 0.001$.

2.2. Biological Activity Assay of NT24

Insecticidal activity experiments on three Lepidoptera pests including *H. armigera*, *S. exigua* and *S. litura* were conducted to assess the difference in the yield and virulence of spinosads between WT and NT24. Compared with the WT strain, NT24 caused higher mortality rates and exhibited stronger insecticidal virulence. Second-stage larvae of *H.armigera*, *S.exigua* and *S.litura* treated with NT24 fermentation broth (0.5 μL/cm^2) for 24 h exhibited more pronounced toxic symptoms, including higher degrees of paralysis and distortion (Figure 2A). Furthermore, the semi-lethal concentration (LC_{50}) of NT24 was significantly reduced (Figure 2B). As shown, the LC_{50} values of the WT strain at 48 h were 1.890 μL/cm^2, 1.262 μL/cm^2 and 2.478 μL/cm^2 against *H. armigera*, *S. exigua* and *S. litura*, respectively, while for NT24, the LC_{50} values were 1.305 μL/cm^2, 0.511 μL/cm^2 and 1.372 μL/cm^2, demonstrating that the virulence of NT24 increased by 30.9%, 59.5% and 44.6%, respectively (Table 1).

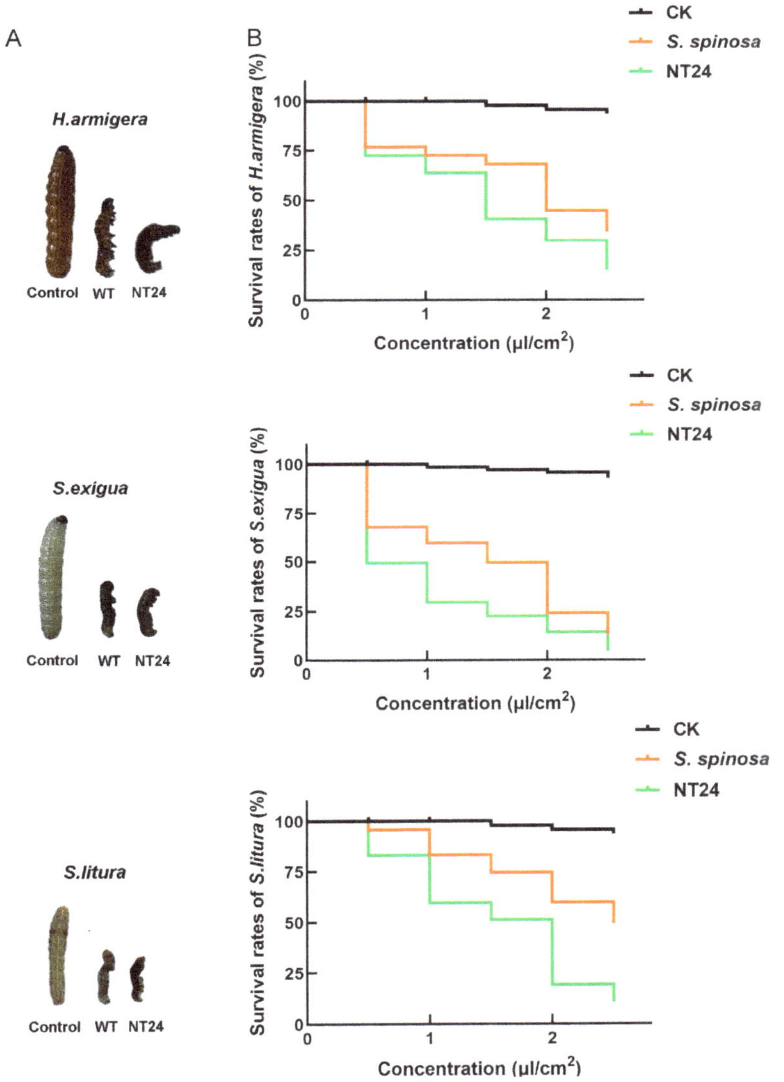

Figure 2. Biological activity comparison of WT and NT24. (**A**) Toxic symptoms of *H. armigera*, *S. exigua* and *S. litura* after treated with the fermentation broth (0.5 μL/cm^2) of WT and NT24 for 48 h. (**B**) The survival rates of *H. armigera*, *S. exigua* and *S. litura* after being treated with different fermentation broths of WT and NT24 for 48 h.

Table 1. LC_{50} analysis of the toxicity of WT and NT24 against *H. armigera*, *S. exigua*, and *S. litura*.

Strains	LC_{50} (μL/cm^2)			95% Confidence Interval		
	Pest					
	H. armigera	*S. exigua*	*S. litura*	*H. armigera*	*S. exigua*	*S. litura*
S. spinosa	1.890	1.262	2.478	1.631–2.259	1.057–1.439	2.176–3.028
NT24	1.305	0.511	1.372	1.053–1.521	0.092–0.768	1.195–1.539

2.3. Phenotypic Differences Between WT and NT24

Given that the growth and metabolic capacity of strains often affect the biosynthesis of secondary metabolites, the determination of glucose consumption and biomass was employed to preliminarily assess the differences between the WT strain and NT24. The results indicate that the glucose consumption rate of NT24 exceeded that of WT from day 2, and its residual glucose level was only 2.8 g/L by day 8 (Figure 3A). The results of biomass determination show that the highest biomass of NT24 reached 29 g/L, which is significantly higher than that of WT, and the growth rate of NT24 was notably faster (Figure 3B). Furthermore, NT24 entered the decline phase on day 6, which was two days earlier than WT.

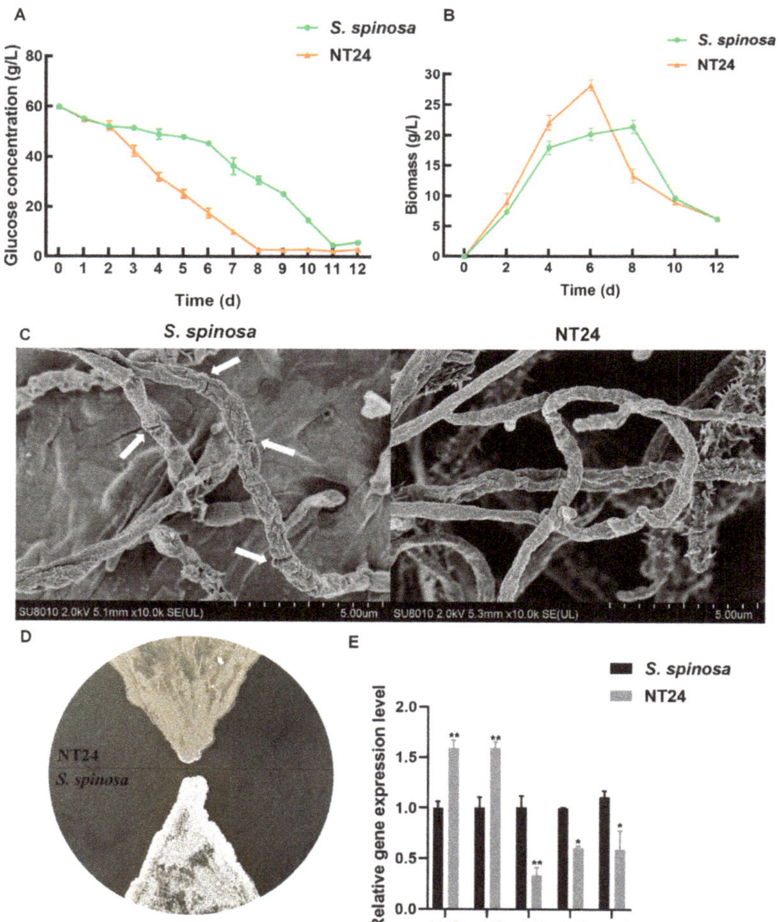

Figure 3. Effects of ARTP/NTG compound mutagenesis on strain growth and development. (**A**) Glucose consumption curve of WT and NT24. (**B**) Biomass accumulation curves of WT and NT24. (**C**) Mycelium comparison of WT and NT24 via SEM. The white arrows point to the differentiation nodes in WT. (**D**) Sporulation capacity comparison of WT and NT24. (**E**) Expression levels of genes related to sporulation and mycelium growth. Error bars are calculated from three independent determinations of mRNA abundance in each sample. Univariate variance, * $p < 0.05$, ** $p < 0.005$.

A comparison of spores and mycelium between the WT strain and NT24 was also conducted. The hyphal morphology during logarithmic growth (3 days) was observed via SEM. On day 3, the differentiation from mycelium to spores was observed in the WT strain, while the NT24 remained in the mycelial state (Figure 3C). The WT strain began to produce spores on day 4, while sporulation in NT24 was delayed until day 6 (Figure 3D). To elucidate this phenomenon, we determined the transcriptional levels of genes involved in mycelial development, including *bldD*, *wblE*, *ssgA*, *whiA*, and *whiB*. RT-qPCR (real-time quantitative polymerase chain reaction) analysis has shown that sporulation-related genes *bldD* and *wblE* were significantly upregulated, which may be the direct cause of the morphological differences between NT24 and the WT strain (Figure 3E).

2.4. Differences of the Accumulation in Spinosad and Short-Chain Acyl-coA

The spinosad gene cluster consists of 23 genes, 19 of which are *spn* genes (*spnA*-*spnS*) associated with forosamine biosynthesis, polyketide briding, tri-O-methylrhamnose biosynthesis and polyketide biosynthesis (Figure 4A). It is speculated that the increase in spinosad production in NT24 may be related to the activation of its spinosad gene cluster. Therefore, 12 *spn* genes were selected based on their transcription direction analyzed by RT-qPCR to evaluate the expression levels of the spinosad gene cluster (Figure 4B). It is obvious that almost all selected *spn* genes in NT24 were upregulated, suggesting that the increased spinosad production in NT24 is primarily mediated by the *spn* genes.

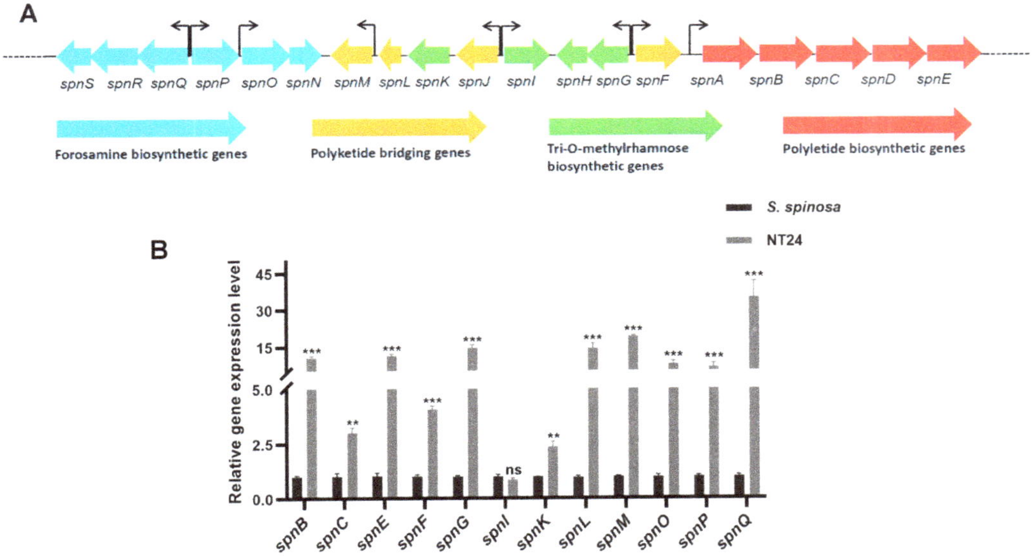

Figure 4. Transcription analysis of the spinosad biosynthetic gene cluster analyzed via RT-qPCR. (**A**) Main structure of the *spn* gene cluster and the transcription direction of each gene. (**B**) Transcript levels of representative *spn* genes. The error bars indicate the standard deviations of three biological replicates. Univariate variance, ^{ns} $p > 0.05$, ** $p < 0.005$, *** $p < 0.001$.

Additionally, the supply of precursors plays a crucial role in the biosynthesis of spinosad. Therefore, the cumulative differences of spinosad production were analyzed in conjunction with the cumulative differences of its short-chain acyl-coA precursors. As shown, NT24 produced a substantial amount of spinosad by day 4 of fermentation, accumulating rapidly, with a maximum titer of 858.3 ± 27.7 mg/L on day 8, whereas the spinosad accumulation rate in WT was considerably slower (Figure 5A). The measurement of short-chain acyl-CoA precursors also revealed that the levels of acetyl-CoA and malonyl-CoA in NT24 increased significantly in all three periods (2d, 4d, 8d), and methylmalonyl-CoA increased significantly on day 4 (Figure 5B–D). These results indicate that the accumulation of acetyl-CoA and malonyl-CoA in NT24 influences its biosynthesis of spinosad. Moreover, malonyl-CoA and acetyl-CoA can be interconverted and stimulate acetyl-CoA synthesis, promoting the TCA cycle and thereby enhancing the growth activity of the strains.

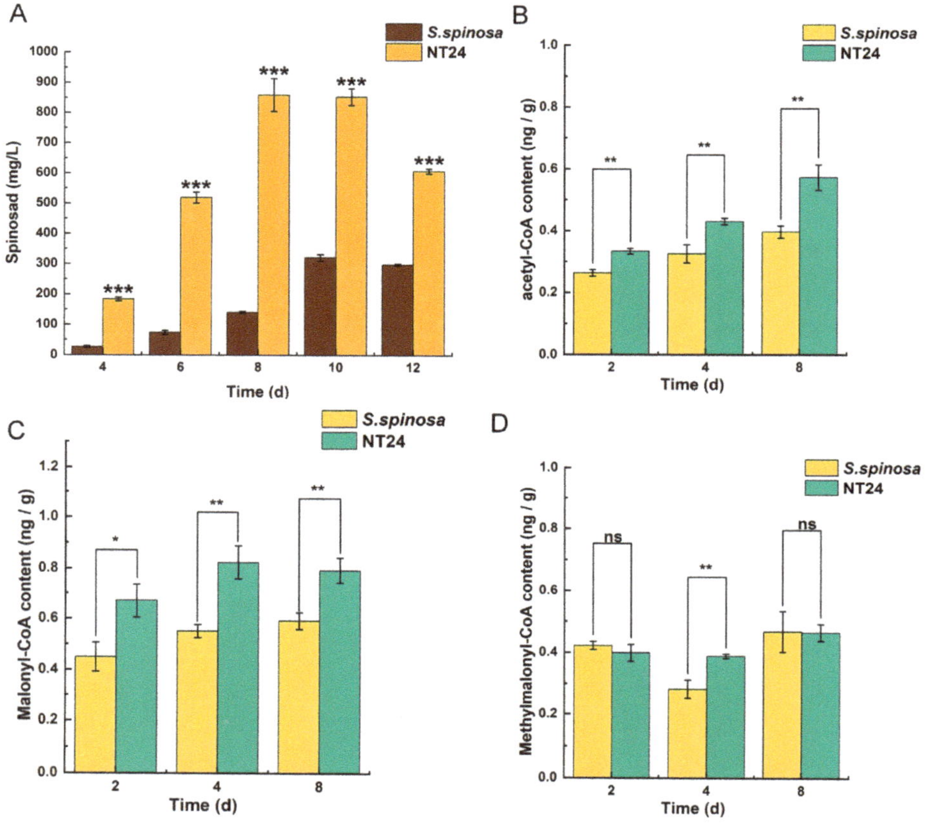

Figure 5. Effects of ARTP/NTG compound mutagenesis on the accumulation of spinosad and acyl-coA pool. (**A**) Spinosad accumulation of WT and NT24. (**B**) Acetyl-CoA accumulation of WT and NT24. (**C**) Malonyl-CoA accumulation of WT and NT24. (**D**) Methylmalonyl-CoA accumulation of WT and NT24. Univariate variance, ^{ns} $p > 0.05$, * $p < 0.05$, ** $p < 0.005$, *** $p < 0.001$.

2.5. Analyses of Resequencing and Metabolomics

Mutations are generally the primary cause of new phenotypes. Therefore, resequencing was performed on NT24 using the whole genome sequence of *S. spinosa* CCTCC M206084 as the reference. A total of 57 SNPs were identified in NT24, of which 48 were located in the CDS region and 30 caused non-synonymous mutations (Table S1). A total of 163 Indels were identified, 67 of which were located in the CDS, and 47 resulted in frameshift mutations (Table S2). The coding sequences of 49 genes were mutated, involving 30 enzymes, 13 proteins related to signaling and cellular processes, two regulators, one transposase and three hypothetical proteins. These mutations are enriched in signaling and cellular processes, carbohydrate metabolism, terpenoid and polyketides metabolism, amino acid metabolism, lipid metabolism and cell growth, purine and pyrimidine metabolism, pyruvate metabolism, and DNA replication and repair, according to the Kyoto Encyclopedia of Genes and Genomes (KEGG). Additionally, two SNPs and 50 Indels were found in the intergenic region of NT24.

Notably, SNPs were found in two genes directly related to pyruvate metabolism, *oadA* (oxaloacetate decarboxylase) and *poxB* (pyruvate oxidase). OadA catalyzes the irreversible decarboxylation of oxaloacetate to pyruvate, while PoxB is responsible for catalyzing the irreversible oxidation of pyruvate to acetate [38,39]. As an important intermediate metabolite, pyruvate can be converted into various precursors required for the biosynthesis of secondary metabolites, including spinosad. Therefore, we hypothesized that the mutations described above may lead to the accumulation of pyruvate in NT24, subsequently resulting in increased spinosad production.

Consequently, targeted metabolomics was employed to analyze differences in metabolite abundance associated with strain growth and target product biosynthesis by LC-MS/MS. The results show that among the 61 kinds of intracellular metabolites identified and quantified by LC-MS/MS, 34 exhibited significant differences, primarily including amino acids, carbohydrates, nucleotides and organic acids and their derivatives, with pyruvate abundance increasing significantly (Table S3). According to KEGG classification, the differential metabolites were mainly enriched in pantothenate and CoA biosynthesis, pyruvate metabolism, amino acid metabolism, the citrate cycle, and glyoxylate and dicarboxylate metabolism (Figure 6A).

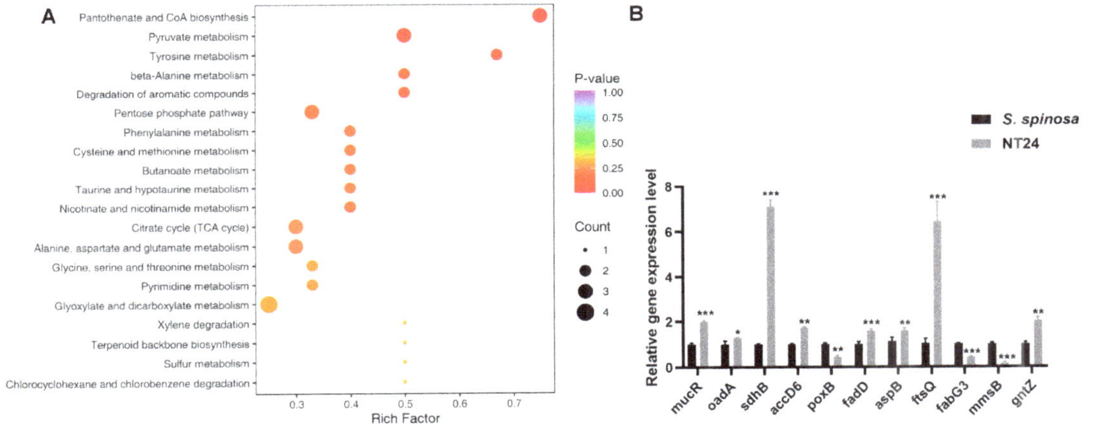

Figure 6. Representative differential genes and KEGG enrichment pathways in NT24. (**A**) Bubble plot of the most significant pathways with KEGG enrichment of DEGs. (**B**) Transcription levels of differential genes in NT24. Univariate variance, * $p < 0.05$, ** $p < 0.005$, *** $p < 0.001$.

RT-qPCR analysis of the mutated genes mentioned above was performed on WT and NT24 to verify the results of resequencing and metabolomics at the transcriptional level (Figure 6B). Differential gene analysis showed that the genes related with carbohydrate metabolism, such as *gntZ*, *sdhB*, *accD*, *oadA* and *fadD*, were upregulated, while *mmsB* and *poxB* were downregulated. In addition, the relative transcription levels of *aspB* (coding aspartate aminotransferase) and *ftsQ* (cell division protein FtsQ) were downregulated.

3. Discussion

Natural products are the result of a long-term evolutionary process of microbial adaptation to the environment, and are closely related to a strain's growth and metabolism [40]. Exploring key genes and their effects on strain growth and secondary metabolism will help to reveal the regulatory mechanism of spinosad biosynthesis and provide guidance for the optimization of other polyketides' production (Figure 7). As shown by biomass measurements, the biomass of NT24 was much higher than that of WT, which may be one of the reasons for the increased yield of spinosad in NT24. *ftsQ* is one of the differential genes in NT24. At the end of a highly conserved cell wall division operon, the *ftsQ* gene encodes a cell division protein that plays an important role in regulating cell growth and death and maintaining cell length, which is necessary for chromosome separation and eventual cell division [41]. The increasing biomass in NT24 may be related to the upregulation of *ftsQ*, which indirectly leads to the increase in spinosad. As for the morphological observation, we found that the sporulation of NT24 was delayed, and SEM observation showed more clearly that the mycelium of NT24 did not show differentiation at day 3. Therefore, the genes related to sporulation and mycelial growth were verified by RT-qPCR, among which the global regulator *bldD* was upregulated 1.6-fold. BldD usually acts as a developmental repressor to control the morphological development of *Streptomyces*, and can also directly regulate the biosynthesis of secondary metabolites [42–44]. In addition, we found an SNP in the promoter region of *mucR*, the diguanylate cyclase coding gene, and the transcriptional level of *mucR* in NT24 was also upregulated. According to previous studies, mutations in a limited number of bases in a promoter can induce significant changes in protein expression [39,45]. Diguanylate cyclase catalyzes GTP cyclization to form c-di-GMP, a signaling molecule with diverse functions, which can regulate the activity of *bldD* and bind with *bldD* to cope with stress or nutrient starvation, and inhibit the differentiation from mycelium to spore as well [46]. Furthermore, c-di-GMP has been verified to participate in the cascade regulation of secondary metabolite biosynthesis [47,48], but its regulation of spinosad biosynthesis remains to be further studied.

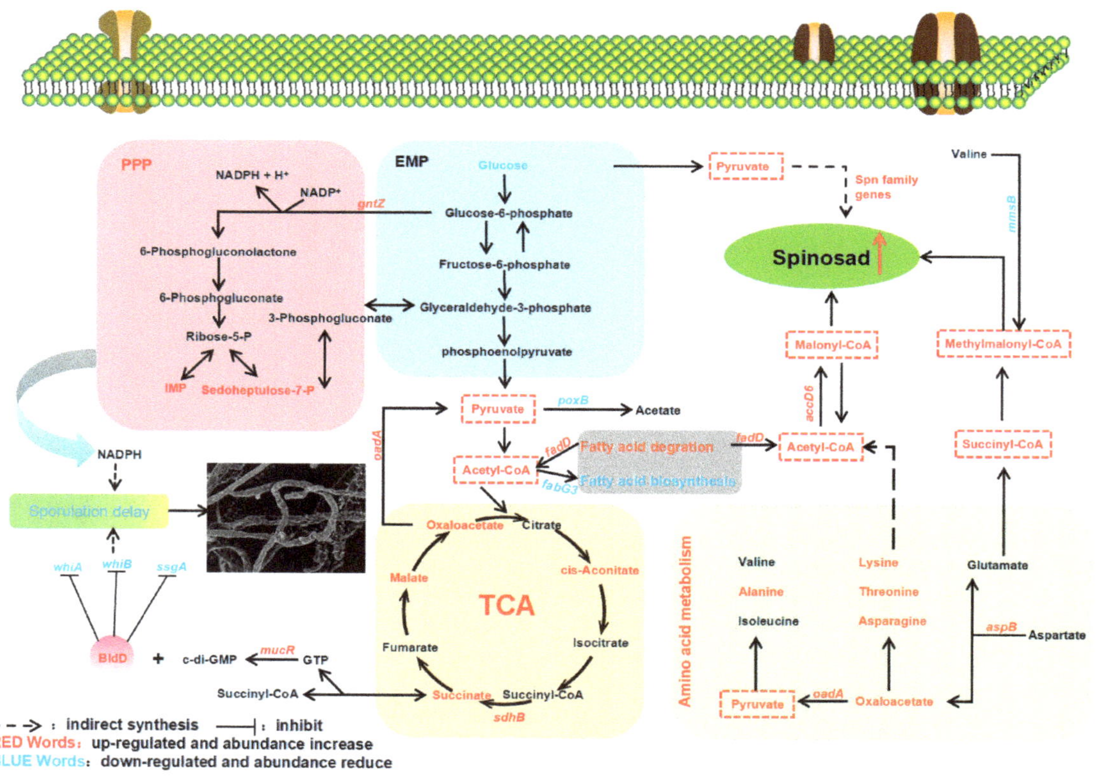

Figure 7. Effects of ARTP/NTG compound mutagenesis on primary and secondary metabolisms of NT24. The pyruvate and acyl-CoA pools are marked with a red box.

The results of spinosad accumulation show that the fermentation cycle of NT24 was shortened by two days, reaching the highest titer of 858.3 ± 27.7 mg/L on the 8th day, which is consistent with the growth trend. The glucose consumption assay showed an accelerated rate of glucose consumption in NT24, which may be related to the previously mentioned acceleration of cell division. Accelerated cell division resulted in more cells in a short period, thus consuming more glucose. In addition, the RT-qPCR results show that the 6-phosphogluconate dehydrogenase (an important enzyme in the pentose phosphate pathway, PPP) coding gene *gntZ* was significantly upregulated in NT24, which catalyzes the generation of ribulose from glucose-6-phosphate. Ribulose can further be converted into sedoheptulose-7-phosphate or IMP. Targeted metabolomics also confirmed significantly increased levels of sedoheptulose-7-phosphate and IMP, suggesting that the upregulation of *gntZ* led to an increase in PPP pathway flux. Meanwhile, PPP is an important source of NADPH, and according to recent analyses of biological intracellular metabolite levels, the balance between NADH production and NADPH consumption by redox metabolism affects the biosynthesis and morphological differentiation of secondary metabolites [49,50].

Spn family genes are key regulatory genes of spinosad biosynthesis, directly affecting the efficiency of its production. The RT-qPCR results showed that almost all selected *spn* genes were upregulated in NT24, which is the direct reason for the improved spinosad production. Moreover, the accumulation of short-chain acyl-CoA precursors is also an important reason for the increase in spinosad. Targeted metabolomics showed that pyruvate content in NT24 increased significantly, and the conversion of pyruvate could alter the contents of propionyl-CoA, malonyl-CoA and methylmalonyl-CoA, ultimately affecting spinosad biosynthesis [51,52]. Therefore, we conclude that pyruvate accumulation indirectly improved the spinosad biosynthesis pathway. Combined with the transcriptional-level analysis of differential genes identified in resequencing and key genes in corresponding metabolic pathways, we found a large metabolic flow into pyruvate and short-chain acyl-coA precursors. RT-qPCR showed that *aspB*, *sdhB*, *accD6*, and *fadD* were significantly upregulated, while *poxB*, *fabG3*, and *mmsB* were significantly downregulated. Aspartate and alpha-ketoglutaric acid can be catalyzed by aspartate aminotransferase (*aspB*) to form oxaloacetate and glutamic acid. Then, oxaloacetic acid can be converted to pyruvate under the catalysis of oxaloacetate decarboxylase (*oadA*) in the glyoxylate cycle, and glutamic acid can be converted to succinyl-CoA, the precursor of spinosad biosynthesis. In NT24, the abundances of malate, succinate, oxaloacetate, and cis-aconitate in the TCA cycle were significantly increased, and the succinate dehydrogenase coding gene *sdhB* was significantly upregulated, indicating the enhancement of the TCA cycle, which might be one of the reasons for the enhanced growth activity of NT24. Long-chain acyl-CoA synthetase (*fadD*) and short-chain dehydrogenase (*fabG3*) are two important rate-limiting enzymes in fatty acid degradation and biosynthesis, respectively. The upregulation of *fadD* and downregulation of *fabG3* enhance fatty acid degradation, which produces a large amount of acetyl-CoA, reducing the competition for acetyl-CoA in fatty acid biosynthesis. In addition, acetyl-CoA can be catalyzed by acetyl-CoA carboxylase (*accD6*) and converted into malonyl-CoA, the direct precursor of spinosad biosynthesis, which is consistent with the above results. The downregulation of *mmsB* (encoding 3-hydroxyisobutyrate dehydrogenase) partly explains why the increasing level of methylmalonyl-CoA was not as high as the levels of acetyl-CoA and malonyl-CoA.

4. Materials and Methods

4.1. Strains and Culture Conditions

The wild-type strain *S. spinosa* CCTCC M206084 (WT, GenBank accession no. CP061007) used in this study was stored in our lab. All the strains used in this study were listed in Table S4. The CSM medium (10 g/L glucose, 45 g/L trypticase soy broth, 9 g/L yeast extract, and 2.2 g/L $MgSO_4 \cdot 7H_2O$) was used for the seed culture in a 100 mL flask, with a starting volume of 20 mL at 30 °C and 260 rpm. After 48 h of cultivation, 1.5 mL activated bacterial solution was inoculated into 30 mL fermentation media (60 g/L glucose; 22.5 g/L cottonseed cake meal, 5 g/L bean cake powder, 2 g/L yeast extract, 10 g/L soluble starch, 7 g/L corn syrup, 5 g/L $CaCO_3$, pH = 7.5) and incubated at 30 °C at 260 rpm, for 10–12 days. All the media were autoclaved at 115 °C for 30 min to confirm the sterility of each batch of broth. Spore morphology was observed in TSB agar plates (tryptic soy broth medium) and incubated at 30 °C for 6 days. All reagents were purchased from Tianheng Biotechnology Co. LTD (Beijing, China).

4.2. Semi-Lethal Concentration Assay of WT and Mutant Strain NT24

The larvae (KEYUN Biology Technology Inc) of *Helicoverpa armigera* (*H. armigera*), *Spodoptera exigua* (*S. exigua*) and *Spodoptera litura* (*S. litura*) were fed with artificial feed at 28 °C with a light/dark cycle of 12 h. Each well of the 24-well cell culture plates (Corning, USA) was added with 0.25 cm^3 (1 cm × 1 cm × 0.25 cm) artificial feed. After the WT and NT24 were fermented for 12 days, the fermentation broth was collected and the artificial feed was sprayed with 0.5, 1, 1.5, 2 and 2.5 ($\mu L/cm^2$) of fermentation broth, ensuring full coverage of the feed surface. The blank control was the corresponding concentration of normal saline. The larvae were cultured in the 24-well cell culture plates (one larva per well) with three replicates. The survival percentage of the larva was recorded after 48 h.

4.3. Detection and Identification of Spinosad

Under the corresponding fermentation time, 3 mL methyl alcohol was added to 1 mL fermentation broth, and the supernatant was centrifuged at 12,000× *g* rpm for 10 min for high-performance liquid chromatography and LC-MS/MS analysis. HPLC (Agilent 1290, wavelength: 250 nm, C18 column: AQ12S05-1546WT, YMC) was used to detect the spinosad titer. Here, 20 μL sample was loaded onto a C18 column (4.5 μm, 4.5 × 150 mm, AQ12S05-1546WT) and detected at 250 nm for 25 min with a rate of 1 mL/min. The mobile phase consisted of acetonitrile (42%), methyl alcohol (42%) and 20 g/L ammonium acetate (16%). The spinosad peak was identified by LC-MS/MS on the LTQ-XL (Thermo Fisher, Waltham, MA, USA) as in the previous study [53].

4.4. ARTP/NTG Mutagenesis

ARTP mutagenesis was performed in the ARTP biological mutagenesis system (Wuxi Yuanqing Tianmu Biological Technology Co., Ltd., Wuxi, China), with a 100 W working radio-frequency power input, 2.0 mm treatment distance and 10 standard liters per minute (SLPM) gas flow. To determine the optimal treatment time, the WT strain's spore suspension (107–108 CFU/mL) was treated in a range of 0–150 s, and the untreated spore suspension was used as control. The treatment time that resulted in a 90% mortality of the strain was selected as the optimal treat duration. The resuspend cell broth was diluted, spread onto TSB agar plates and incubated in 30 °C for 6 days. Each clone on these plates was activated on 2 mL CSM medium for 48 h, and a sample from 3 random sights was selected to observe the mycelium; the mutants containing stronger mycelium were screened out and cultured in the fermentation media at 30 °C for 12 days.

HPLC was used to detect the spinosad titer. The mutant strain with the highest spinosad titer in ARTP mutagenesis was used as the original strain in NTG mutagenesis. And the optimum NTG concentration was selected from the lethality rates (reach 90%) under different NTG concentrations (0, 100, 200, 300, 400, and 500 μg/mL) for 1 h. The follow-up steps here were the same as in ARTP mutagenesis.

4.5. Growth Curve and Glucose Consumption Determination

During fermentation, 5 mL of fermentation broth was collected from the shake flask every 2 days, filtered through three layers of pre-weighed filter paper, and placed in a 50 °C oven. After 12 h, the sample was removed and re-weighed. The growth curve was measured over 12 days with three biological replicates. To determine the glucose content, 1.5 mL dinitrosalicylic acid reagent was added to 1 mL of fermentation broth supernatant daily during the 12-day fermentation with three biological replicates. The mixture was incubated in a water bath at 100 °C for 2 min. After adding ultrapure water to a final volume of 10 mL, the absorbance was measured (540 nm), and glucose content was calculated based on the dilution factor and optical density (OD) value.

4.6. Morphological Observation

The mutants activated in CSM medium for 48 h were observed with a phase contrast microscope (AXIO Scope A1, Zeiss, Germany). WT and NT24 were activated in CSM medium for 48 h, and 20 μL of each was taken out and spread on TSB solid medium. After that, the sterilized cover glass was inserted into TSB solid medium aslant. After 3 days of culture at 30 °C, the cover glass was taken out and dried for sputtering gold plating, and then imaged with a scanning electron microscope (SEM, Hitachi, Shinagawa-ku SU8010, Japan).

4.7. Determination of Short-Chain Acyl-CoA Precursors

Biomass samples of WT an NT24 were harvested on the 2nd, 4th, and 8th days, and sufficiently broken by grinding with liquid nitrogen to release intracellular components. The supernatants were collected (8000 rpm, 10 min) and analyzed according to the manufacturers' instructions using microorganism acetyl-CoA/malonyl-CoA/methyl-malonyl-CoA ELISA Kits (Jiangsu JINMEI Biotechnology Co., Ltd., Jiangsu, China). Three biological replicates were conducted in this experiment.

4.8. Genome Sequencing and Assembly

The genome of NT24 was sequenced using an DNBSEQ platform at the Beijing Genomics Institute (Shenzhen, China). Genomic DNA was sheared randomly to construct three read libraries with lengths of 8738,130 bp via physico-chemical methods. The paired-end fragment libraries were sequenced. Raw reads of low quality produced by paired-end sequencing (those with consecutive bases covered by fewer than five reads) were discarded. The sequenced reads were assembled using SOAPdenovo v1.05 software. All mentioned metabolites were detected by MetWare (http://www.metware.cn/) (accessed on 2 February 2024) based on the AB Sciex QTRAP 6500 LC-MS/MS platform.

4.9. Total RNA Extraction and RT-qPCR Analysis

The total RNA of WT and NT24 at 96 h was extracted following the instructions for the total RNA Extractor (Sangon, Shanghai, China). NanoDrop 2000 (Thermo) was used to test the purity and concentration of extracted RNA. The total RNA was reverse-transcribed into cDNA using a RevertAidTM First Strand cDNA Synthesis Kit (Fermentas, Waltham, MA, USA) following the instructions. All samples were analyzed on a 7500 Real-Time PCR system instruments (Applied Biosystems, Norwalk, CT, USA). The 16S rRNA gene was regarded as the endogenous control to quantify the relative expression levels of the target genes. The RT-qPCR primers used in this study are listed in Table S5. The experiment was repeated 3 times.

4.10. Stability Analysis

Genetic stability testing was conducted by subculturing NT24 in 30 mL of fermentation medium for five passages (the transduction was performed once every 8 days), and each passage was subjected to HPLC analysis.

4.11. Statistic Method

The mortality of larvae (%) = (A/72) × 100%, where A is the number of dead larvae counted across three 24-well plates. The LC_{50} and statistical significance analysis were performed using the SPSS software (Inc, version 20, Chicago, IL, USA). The lethality rate of ARTP/NTG mutagenesis (%) = ((A − B)/A) × 100%, where A is the number of control colonies on TSB plate, and B is the number of ARTP/NTG mutagenesis colonies at the same gradient.

5. Conclusions

In conclusion, a mutant strain NT24 with enhanced spinosad biosynthesis was obtained through ARTP/NTG compound mutagenesis in this study, where the accumulation of pyruvate and acyl-CoA precursors plays a central role in spinosad production. The four key genes (*oadA*, *poxB*, *accD6* and *fadD*), identified through metabolomics and RT-qPCR, could be overexpressed or knocked out/down in future studies to validate their functions and effects on spinosad biosynthesis. In addition, in contrast to genetically engineered strains (mutant strains obtained through genetic modification), the mutant strains obtained via ARTP/NTG compound mutagenesis circumvent the regulatory restrictions on genetically modified products in agricultural applications, offering greater versatility. Moreover, the mutant strain generated from mutagenesis can serve as an "original strain" for further genetic modifications to enhance its spinosad titer. The method adopted in this study can also be applied to optimize the biosynthesis of secondary metabolites in other actinomycetes.

Supplementary Materials: The following supporting information can be downloaded at https://www.mdpi.com/article/10.3390/ijms252212308/s1.

Author Contributions: Conceptualization, L.X.; funding acquisition, J.R. and L.X.; investigation, Z.Z., W.C. and Z.X.; methodology, Z.Z. and L.X.; software, Z.Z., W.C. and Z.X.; supervision, S.H. and L.X.; visualization, Z.Z., W.C. and Z.X.; writing—original draft, Z.Z.; writing—review and editing, W.C. and L.C. All authors have read and agreed to the published version of the manuscript.

Funding: This work was supported by funding from the National Natural Science Foundation of China (31770106, 32200062) and the Natural Science Foundation of Hunan Province (2024JJ5258).

Institutional Review Board Statement: Not applicable.

Informed Consent Statement: Not applicable.

Data Availability Statement: The raw data supporting the conclusions of this article will be made available by the authors on request.

Conflicts of Interest: The authors declare no conflicts of interest.

References

1. Al-Fadhli, A.A.; Threadgill, M.D.; Mohammed, F.; Sibley, P.; Al-Ariqi, W.; Parveen, I. Macrolides from rare actinomycetes: Structures and bioactivities. *Int. J. Antimicrob. Agents* **2022**, *59*, 106523. [CrossRef] [PubMed]
2. Ding, T.; Yang, L.-J.; Zhang, W.-D.; Shen, Y.-H. The secondary metabolites of rare actinomycetes: Chemistry and bioactivity. *RSC Adv.* **2019**, *9*, 21964–21988. [CrossRef] [PubMed]
3. Tiwari, K.; Gupta, R.K. Rare actinomycetes: A potential storehouse for novel antibiotics. *Crit. Rev. Biotechnol.* **2011**, *32*, 108–132. [CrossRef] [PubMed]
4. Guojun, Y.; Yuping, H.; Yan, J.; Kaichun, L.; Haiyang, X. A New Medium for Improving Spinosad Production by *Saccharopolyspora spinosa*. *Jundishapur J. Microbiol.* **2016**, *9*, e16765. [CrossRef]
5. Newman, D.J.; Cragg, G.M. Natural Products as Sources of New Drugs from 1981 to 2014. *J. Nat. Prod.* **2016**, *79*, 629–661. [CrossRef]
6. Salgado, V.L.; Sparks, T.C.; Gilbert, L.I.; Gill, S.S. The Spinosyns: Chemistry, Biochemistry, Mode of Action, and Resistance. In *Insect Control: Biological and Synthetic Agents*; Academic Press: Cambridge, MA, USA, 2010; pp. 137–169.
7. Madduri, K.; Waldron, C.; Merlo, D.J. Rhamnose Biosynthesis Pathway Supplies Precursors for Primary and Secondary Metabolism in *Saccharopolyspora spinosa*. *J. Bacteriol.* **2001**, *183*, 5632–5638. [CrossRef]
8. Huang, K.-x.; Xia, L.; Zhang, Y.; Ding, X.; Zahn, J.A. Recent advances in the biochemistry of spinosyns. *Appl. Microbiol. Biotechnol.* **2009**, *82*, 13–23. [CrossRef]
9. Kirst, H.A. The spinosyn family of insecticides: Realizing the potential of natural products research. *J. Antibiot.* **2010**, *63*, 101–111. [CrossRef]
10. Santos, V.S.V.; Pereira, B.B. Properties, toxicity and current applications of the biolarvicide spinosad. *J. Toxicol. Environ. Health B Crit. Rev.* **2020**, *23*, 13–26. [CrossRef]
11. Tamilselvan, R.; Kennedy, J.S.; Suganthi, A. Sublethal and transgenerational effects of spinetoram on the biological traits of *Plutella xylostella* (L.) (Lepidoptera: Plutellidae). *Ecotoxicology* **2021**, *30*, 667–677. [CrossRef]
12. Nault, B.A.; Sandhi, R.K.; Harding, R.S.; Grundberg, E.A.; Rusinek, T.; Munyaneza, J. Optimizing Spinosyn Insecticide Applications for *Allium Leafminer* (Diptera: Agromyzidae) Management in Allium Crops. *J. Econ. Entomol.* **2022**, *115*, 618–623. [CrossRef] [PubMed]

13. Tang, L.-D.; Guo, L.-H.; Ali, A.; Desneux, N.; Zang, L.-S.; Musser, F. Synergism of Adjuvants Mixed With Spinetoram for the Management of Bean Flower Thrips,*Megalurothrips usitatus* (Thysanoptera: Thripidae) in Cowpeas. *J. Econ. Entomol.* **2022**, *115*, 2013–2019. [CrossRef] [PubMed]
14. Mayes, M.A.; Thompson, G.D.; Husband, B.; Miles, M.M. Spinosad Toxicity to Pollinators and Associated Risk. *Rev. Environ. Contam. Toxicol.* **2002**, *179*, 37–71. [CrossRef]
15. Sparks, T.C.; Crouse, G.D.; Durst, G. Natural products as insecticides: The biology, biochemistry and quantitative structure–activity relationships of spinosyns and spinosoids. *Pest Manag. Sci.* **2001**, *57*, 896–905. [CrossRef]
16. Perry, T.; Chen, W.; Ghazali, R.; Yang, Y.T.; Christesen, D.; Martelli, F.; Lumb, C.; Bao Luong, H.N.; Mitchell, J.; Holien, J.K.; et al. Role of nicotinic acetylcholine receptor subunits in the mode of action of neonicotinoid, sulfoximine and spinosyn insecticides in *Drosophila melanogaster*. *Insect Biochem. Mol. Biol.* **2021**, *131*, 103547. [CrossRef]
17. Yang, M.; Wang, B.; Gao, J.; Zhang, Y.; Xu, W.; Tao, L. Spinosad induces programmed cell death involves mitochondrial dysfunction and cytochrome C release in *Spodoptera frugiperda* Sf9 cells. *Chemosphere* **2017**, *169*, 155–161. [CrossRef]
18. Gokbulut, C.; Ozuicli, M.; Aslan, B.; Aydin, L.; Cirak, V.Y. The residue levels of spinosad and abamectin in eggs and tissues of laying hens following spray application. *Avian Pathol.* **2019**, *48* (Suppl. S1), S44–S51. [CrossRef]
19. Liu, Y.; Sun, H.; Wang, S. Dissipation and Residue of Spinosad in *Zucchini* Under Field Conditions. *Bull. Environ. Contam. Toxicol.* **2013**, *91*, 256–259. [CrossRef]
20. Sarfraz, M.; Dosdall, L.M.; Keddie, B.A. Spinosad: A Promising Tool for Integrated Pest Management. *Outlooks Pest Manag.* **2005**, *16*, 78–84. [CrossRef]
21. Tao, H.; Zhang, Y.; Deng, Z.; Liu, T. Strategies for Enhancing the Yield of the Potent Insecticide Spinosad in Actinomycetes. *Biotechnol. J.* **2018**, *14*, e1700769. [CrossRef]
22. Bridget, A.F.; Nguyen, C.T.; Magar, R.T.; Sohng, J.K. Increasing production of spinosad in *Saccharopolyspora spinosa* by metabolic engineering. *Biotechnol. Appl. Biochem.* **2022**, *70*, 1035–1043. [CrossRef] [PubMed]
23. Liu, Z.; Zhu, Z.; Tang, J.; He, H.; Wan, Q.; Luo, Y.; Huang, W.; Yu, Z.; Hu, Y.; Ding, X.; et al. RNA-Seq-Based Transcriptomic Analysis of *Saccharopolyspora spinosa* Revealed the Critical Function of PEP Phosphonomutase in the Replenishment Pathway. *J. Agric. Food Chem.* **2020**, *68*, 14660–14669. [CrossRef] [PubMed]
24. Jha, A.K.; Pokhrel, A.R.; Chaudhary, A.K.; Park, S.-W.; Cho, W.J.; Sohng, J.K. Metabolic Engineering of Rational Screened *Saccharopolyspora spinosa* for the Enhancement of Spinosyns A and D Production. *Mol. Cells* **2014**, *37*, 727–733. [CrossRef]
25. Ottenheim, C.; Nawrath, M.; Wu, J.C. Microbial mutagenesis by atmospheric and room-temperature plasma (ARTP): The latest development. *Bioresour. Bioprocess.* **2018**, *5*, 12. [CrossRef]
26. Buyuklyan, J.A.; Zakalyukina, Y.V.; Osterman, I.A.; Biryukov, M.V. Modern Approaches to the Genome Editing of Antibiotic Biosynthetic Clusters in Actinomycetes. *Acta Naturae* **2023**, *15*, 4–16. [CrossRef]
27. Zhao, Y.; Li, G.; Chen, Y.; Lu, Y. Challenges and Advances in Genome Editing Technologies in *Streptomyces*. *Biomolecules* **2020**, *10*, 734. [CrossRef]
28. Twardowski, T.; Małyska, A. Uninformed and disinformed society and the GMO market. *Trends Biotechnol.* **2015**, *33*, 1–3. [CrossRef]
29. Zhang, X.; Zhang, C.; Zhou, Q.Q.; Zhang, X.F.; Wang, L.Y.; Chang, H.B.; Li, H.P.; Oda, Y.; Xing, X.H. Quantitative evaluation of DNA damage and mutation rate by atmospheric and room-temperature plasma (ARTP) and conventional mutagenesis. *Appl. Microbiol. Biotechnol.* **2015**, *99*, 5639–5646. [CrossRef]
30. Gu, C.; Wang, G.; Mai, S.; Wu, P.; Wu, J.; Wang, G.; Liu, H.; Zhang, J. ARTP mutation and genome shuffling of ABE fermentation symbiotic system for improvement of butanol production. *Appl. Microbiol. Biot.* **2017**, *101*, 2189–2199. [CrossRef]
31. Maslowska, K.H.; Makiela-Dzbenska, K.; Fijalkowska, I.J. The SOS system: A complex and tightly regulated response to DNA damage. *Environ. Mol. Mutagen.* **2019**, *60*, 368–384. [CrossRef]
32. Wang, L.; Zhao, H.; He, D.; Wu, Y.; Jin, L.; Li, G.; Su, N.; Li, H.; Xing, X.H. Insights into the molecular-level effects of atmospheric and room-temperature plasma on mononucleotides and single-stranded homo- and hetero-oligonucleotides. *Sci. Rep.* **2020**, *10*, 14298. [CrossRef] [PubMed]
33. Zhang, Q.; Miao, R.; Feng, R.; Yan, J.; Wang, T.; Gan, Y.; Zhao, J.; Lin, J.; Gan, B. Application of Atmospheric and Room-Temperature Plasma (ARTP) to Microbial Breeding. *Curr. Issues Mol. Biol.* **2023**, *45*, 6466–6484. [CrossRef] [PubMed]
34. Zhang, X.; Zhang, X.F.; Li, H.P.; Wang, L.Y.; Zhang, C.; Xing, X.H.; Bao, C.Y. Atmospheric and room temperature plasma (ARTP) as a new powerful mutagenesis tool. *Appl. Microbiol. Biotechnol.* **2014**, *98*, 5387–5396. [CrossRef]
35. Xu, Y.; Jing, Y.; Zhang, Q.; Xiu, J.; Tian, M.; Cui, Q.; Ma, Y.; Yi, L.; Han, L.; Qian, Y.; et al. Improving rhamnolipids biosynthesis in *Pseudomonas* sp. L01 through atmospheric and room-temperature plasma (ARTP) mutagenesis. *Microorganisms* **2023**, *11*, 1182. [CrossRef]
36. Yu, L.; Li, F.; Ni, J.; Qin, X.; Lai, J.; Su, X.; Li, Z.; Zhang, M. UV-ARTP compound mutagenesis breeding improves macrolactins production of *Bacillus siamensis* and reveals metabolism changes by proteomic. *J. Biotechnol.* **2024**, *381*, 36–48. [CrossRef]
37. Zhu, Z.; Chen, W.; Zhou, H.; Cheng, H.; Luo, S.; Zhou, K.; Zhou, P.; Xia, L.; Ding, X. ARTP and NTG compound mutations improved Cry protein production and virulence of *Bacillus thuringiensis* X023. *Appl. Microbiol. Biotechnol.* **2022**, *106*, 4211–4221. [CrossRef]
38. Li, D.; Shen, J.; Ding, Q.; Wu, J.; Chen, X. Recent progress of atmospheric and room-temperature plasma as a new and promising mutagenesis technology. *Cell Biochem. Funct.* **2024**, *42*, e3991. [CrossRef]
39. Phan, T.T.; Nguyen, H.D.; Schumann, W. Development of a strong intracellular expression system for *Bacillus subtilis* by optimizing promoter elements. *J. Biotechnol.* **2012**, *157*, 167–172. [CrossRef]

40. Liu, G.; Chater, K.F.; Chandra, G.; Niu, G.; Tan, H. Molecular Regulation of Antibiotic Biosynthesis in *Streptomyces*. *Microbiol. Mol. Biol. Rev.* **2013**, *77*, 112–143. [CrossRef]
41. Jain, P.; Malakar, B.; Khan, M.Z.; Lochab, S.; Singh, A.; Nandicoori, V.K. Delineating FtsQ-mediated regulation of cell division in *Mycobacterium tuberculosis*. *J. Biol. Chem.* **2018**, *293*, 12331–12349. [CrossRef]
42. Li, J.; Wang, N.; Tang, Y.; Cai, X.; Xu, Y.; Liu, R.; Wu, H.; Zhang, B. Developmental regulator BldD directly regulates lincomycin biosynthesis in *Streptomyces lincolnensis*. *Biochem. Biophys. Res. Commun.* **2019**, *518*, 548–553. [CrossRef] [PubMed]
43. Mouri, Y.; Konishi, K.; Fujita, A.; Tezuka, T.; Ohnishi, Y.; DiRita, V.J. Regulation of Sporangium Formation by BldD in the Rare Actinomycete *Actinoplanes missouriensis*. *J. Bacteriol.* **2017**, *199*, e00840-16. [CrossRef] [PubMed]
44. Yan, H.; Lu, X.; Sun, D.; Zhuang, S.; Chen, Q.; Chen, Z.; Li, J.; Wen, Y. BldD, a master developmental repressor, activates antibiotic production in two *Streptomyces* species. *Mol. Microbiol.* **2019**, *113*, 123–142. [CrossRef] [PubMed]
45. Zhou, C.; Ye, B.; Cheng, S.; Zhao, L.; Liu, Y.; Jiang, J.; Yan, X. Promoter engineering enables overproduction of foreign proteins from a single copy expression cassette in *Bacillus subtilis*. *Microb. Cell Fact.* **2019**, *18*, 111. [CrossRef] [PubMed]
46. Fu, Y.; Dong, Y.-Q.; Shen, J.-L.; Yin, B.-C.; Ye, B.-C.; You, D. A meet-up of acetyl phosphate and c-di-GMP modulates BldD activity for development and antibiotic production. *Nucleic Acids Res.* **2023**, *51*, 6870–6882. [CrossRef]
47. Bechthold, A.; Zechel, D.L.; Paululat, T.; Nuzzo, D.; Tsypik, O.; Makitrynskyy, R. Secondary nucleotide messenger c-di-GMP exerts a global control on natural product biosynthesis in *Streptomycetes*. *Nucleic Acids Res.* **2020**, *48*, 1583–1598. [CrossRef]
48. Nuzzo, D.; Makitrynskyy, R.; Tsypik, O.; Bechthold, A. Identification and Characterization of Four c-di-GMP-Metabolizing Enzymes from *Streptomyces ghanaensis* ATCC14672 Involved in the Regulation of Morphogenesis and Moenomycin A Biosynthesis. *Microorganisms* **2021**, *9*, 284. [CrossRef]
49. Craney, A.; Ahmed, S.; Nodwell, J. Towards a new science of secondary metabolism. *J. Antibiot.* **2013**, *66*, 387–400. [CrossRef]
50. Jin, X.-M.; Chang, Y.-K.; Lee, J.H.; Hong, S.-K. Effects of Increased NADPH Concentration by Metabolic Engineering of the Pentose Phosphate Pathway on Antibiotic Production and Sporulation in *Streptomyces lividans* TK24. *J. Microbiol. Biotechnol.* **2017**, *27*, 1867–1876. [CrossRef]
51. Swiercz, J.P.; Nanji, T.; Gloyd, M.; Guarné, A.; Elliot, M.A. A novel nucleoid-associated protein specific to the actinobacteria. *Nucleic Acids Res.* **2013**, *41*, 4171–4184. [CrossRef]
52. Zhao, C.; Huang, Y.; Guo, C.; Yang, B.L.; Zhang, Y.; Lan, Z.; Guan, X.; Song, Y.; Zhang, X.L. Heterologous expression of spinosyn biosynthetic gene cluster in species is dependent on the expression of rhamnose biosynthesis genes. *J. Mol. Microb. Biotech.* **2017**, *27*, 190–198. [CrossRef]
53. Liu, Z.; Xiao, J.; Tang, J.; Liu, Y.; Shuai, L.; Cao, L.; Xia, Z.; Ding, X.; Rang, J.; Xia, L. Effects of *acuC* on the growth development and spinosad biosynthesis of *Saccharopolyspora spinosa*. *Microb. Cell Fact.* **2021**, *20*, 141. [CrossRef]

International Journal of
Molecular Sciences

MDPI

Article

Mining Translation Inhibitors by a Unique Peptidyl-Aminonucleoside Synthetase Reveals Cystocin Biosynthesis and Self-Resistance

Vera A. Alferova [1,†], Polina A. Zotova [2,†], Anna A. Baranova [1], Elena B. Guglya [1], Olga A. Belozerova [1], Sofiya O. Pipiya [1], Arsen M. Kudzhaev [1], Stepan E. Logunov [1], Yuri A. Prokopenko [1], Elisaveta A. Marenkova [1], Valeriya I. Marina [2], Evgenia A. Novikova [2], Ekaterina S. Komarova [3], Irina P. Starodumova [1,4], Olga V. Bueva [4], Lyudmila I. Evtushenko [4], Elena V. Ariskina [4], Sergey I. Kovalchuk [1], Konstantin S. Mineev [1], Vladislav V. Babenko [5], Petr V. Sergiev [2,3,6], Dmitrii A. Lukianov [2,6] and Stanislav S. Terekhov [1,*]

1 Shemyakin-Ovchinnikov Institute of Bioorganic Chemistry, Miklukho-Maklaya 16/10, 117997 Moscow, Russia; alferovava@gmail.com (V.A.A.); anjabaranowa@list.ru (A.A.B.); eguglya@gmail.com (E.B.G.); o.belozyorova@gmail.com (O.A.B.); pipiyasofiya@ibch.ru (S.O.P.); kudzhaev_arsen@mail.ru (A.M.K.); stepa.2003@bk.ru (S.E.L.); tetrahydrofuran@mail.ru (Y.A.P.); marenkova.lizzz@gmail.com (E.A.M.); iri-starodumova@yandex.ru (I.P.S.); xerx222@gmail.com (S.I.K.); mineev@nmr.ru (K.S.M.)

2 Department of Chemistry, Lomonosov Moscow State University, 119992 Moscow, Russia; zotova-polina98@mail.ru (P.A.Z.); ymmo@mail.ru (V.I.M.); zhizhenzha@gmail.com (E.A.N.); petya@genebee.msu.ru (P.V.S.); dmitrii.a.lukianov@gmail.com (D.A.L.)

3 A.N. Belozersky Institute of Physico-Chemical Biology, Lomonosov Moscow State University, 119991 Moscow, Russia; ekaandreyanova@yandex.ru

4 All-Russian Collection of Microorganisms (VKM), Pushchino Scientific Center for Biological Research, Russian Academy of Sciences, 142290 Pushchino, Russia; oboueva@mail.ru (O.V.B.); lie99@mail.ru (L.I.E.); ariskina@pbcras.ru (E.V.A.)

5 Lopukhin Federal Research and Clinical Center of Physical-Chemical Medicine, Malaya Pirogovskaya Str. 1a, 119435 Moscow, Russia; daniorerio34@gmail.com

6 Center for Molecular and Cellular Biology, 121205 Moscow, Russia

* Correspondence: sterekhoff@gmail.com

† These authors contributed equally to this work.

Abstract: Puromycin (Puro) is a natural aminonucleoside antibiotic that inhibits protein synthesis by its incorporation into elongating peptide chains. The unique mechanism of Puro finds diverse applications in molecular biology, including the selection of genetically engineered cell lines, in situ protein synthesis monitoring, and studying ribosome functions. However, the key step of Puro biosynthesis remains enigmatic. In this work, pur6-guided genome mining is carried out to explore the natural diversity of Puro-like antibiotics. The diversity of biosynthetic gene cluster (BGC) architectures suggests the existence of distinct structural analogs of puromycin encoded by pur-like clusters. Moreover, the presence of $tRNA^{Cys}$ in some BGCs, i.e., *cst*-like clusters, leads us to the hypothesis that Pur6 utilizes aminoacylated tRNA as an activated peptidyl precursor, resulting in cysteine-based analogs. Detailed metabolomic analysis of *Streptomyces* sp. VKM Ac-502 containing *cst*-like BGC revealed the production of a cysteinyl-based analog of Puro—cystocin (Cst). Similar to puromycin, cystocin inhibits both prokaryotic and eukaryotic translation by the same mechanism. Aminonucleoside N-acetyltransferase CstC inactivated Cst, mediating antibiotic resistance in genetically modified bacteria and human cells. The substrate specificity of CstC originated from the steric hindrance of its active site. We believe that novel aminonucleosides and their inactivating enzymes can be developed through the directed evolution of the discovered biosynthetic machinery.

Keywords: puromycin; cystocin; aminonucleoside antibiotics; resistance; genome mining; antimicrobial activity; cytotoxicity

1. Introduction

Puromycin (Puro) is a well-known aminonucleoside antibiotic that blocks protein synthesis [1]. Puro mimics the aminoacyl end of tRNA entering the ribosome A site. Puro is transferred to the growing peptide chain, causing the formation of a puromycylated nascent chain and premature chain release [2]. This mode of action makes Puro a valuable tool for various applications in molecular biology research. For example, puromycylation, typically detected by anti-puromycin antibodies, is used for monitoring protein synthesis in living cells [3,4] and to study the subcellular localization of translating ribosomes [5]. Puro-induced release of ribosome-bound nascent chains is employed in various studies of ribosome functions, including ribosome profiling [5,6] and in situ translation [7,8]. Additionally, puromycin coupling is a crucial step for mRNA display technology [9].

Puro was isolated from *Streptomyces alboniger*, and its biosynthesis pathway attracted significant attention [10–22]. After initial identification of the *pur* biosynthetic gene cluster (BGC) [11], individual enzymes were studied in detail, including NAD-dependent ATP dehydrogenase Pur10 [10], tyrosinyl-aminonucleoside synthetase Pur6 [12], monophosphatase Pur3 [21], O-methyltransferase [17], and resistance-mediating enzymes Pur8 [13], PAC [14,22], Pur7 [20], and N-acetylpuromycin N-acetylhydrolase [15]. Moreover, the regulation mechanism of Puro production was investigated [16,18,19]. The self-resistance mechanism involving the acetylation of the antibiotic into inactive form by the puromycin N-acetyltransferase (PAC) [22] found practical application. This mechanism underlies a widely used cell line selection system [5,23]. Despite its widespread use in biotechnology, the key step in puromycin biosynthesis remains elusive [11]. In this work, we investigate the diversity of Puro-like BGCs to reveal the chemical space of Puro analogs and biosynthetic mechanisms behind aminonucleoside assembly.

2. Results

2.1. Genome Mining for Puromycin-like Antibacterials

The BGC of puromycin provides clues for puromycin biosynthesis. However, we are still far from a detailed mechanism of it. The most uncommon step in puromycin biosynthesis is the coupling of amino acid tyrosine and nucleotide moieties (Figure 1A). According to puromycin BGC, this step must be mediated by a unique enzyme, Pur6 [11]. Pur6 is particularly interesting since this 84 kDa protein has no known conserved domains that could be identified by its sequence homology. Moreover, its AlfaFold 3 structure model has only weak and partial structural homology with cysteine proteinase gingipain R, which is the most proximate structural homolog known yet (Figure 1B).

RMSD between the 139–404 fragment of the Pur6 model and the 69–349 fragment of gingipain R is 2.14 Å. The Cys326 residue of Pur6 is located close to the active Cys244 residue that mediates the proteolytic activity of gingipain R, which indicates its potential involvement in the enzymatic reaction. However, the majority of Pur6 structures could not be assessed even by structural homology, which is outstanding for such a large enzyme.

To understand the diversity and potential catalytic mechanism of *pur6*-like enzymes, we performed genome mining of sequences similar to *pur6* (Figure 2). Pur6-like enzymes were strongly associated with puromycin BGCs in various *Streptomyces* species, including *S. alboniger*, *S. cacaoi*, *S. luteocolor*, *S. silaceus*, *S. harbinensis*, *S. megasporus*, and also in *Saccharopolyspora phatthalungensis*. While transcription regulation and transport mechanisms were often different, the core BGC architecture was generally the same (Figure 2A). Instead, an alternative BGC architecture and methyltransferase were detected for *S. monomycini*, *S. paromomycinus*, and *S. albus*. Moreover, this architecture was detected in *Nocardiopsis kunsanensis*, a member of a different actinomycete order, *Propionibacteriales*, and even in the gammaproteobacterium *Xenorhabdus bovienii*, which belongs to the phylogenetic branch very distant from actinomycetes. This may indicate both an alternative biosynthetic product and potential horizontal transfer between these species.

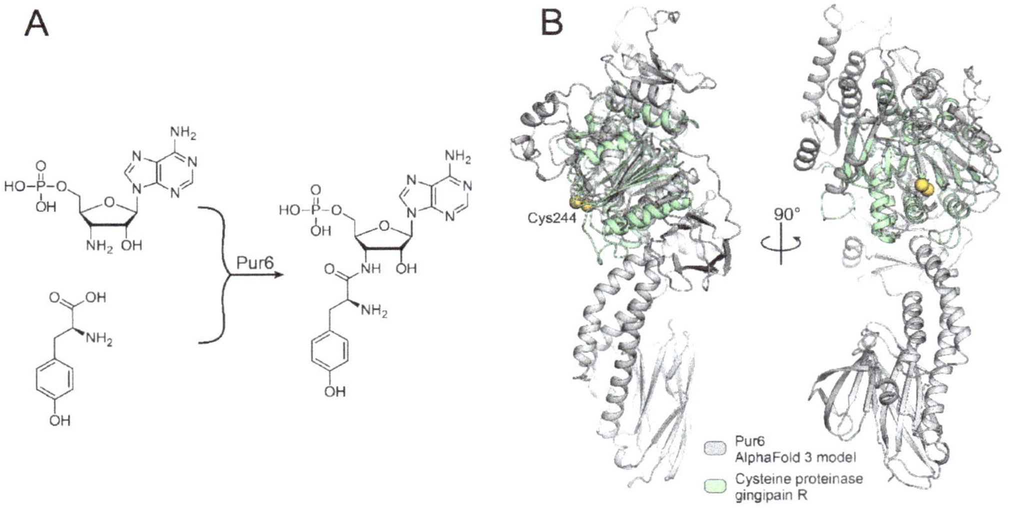

Figure 1. The key step of puromycin biosynthesis is mediated by the unique enzyme Pur6. (**A**) A schematic representation of the reaction catalyzed by Pur6 that was proposed in [11]. (**B**) Pur6 structure model generated by AlfaFold3 (grey) aligned with cysteine proteinase (green) gingipain R (PDB ID: 1CVR). Essential catalytic Cys244 residue of gingipain R and Cys326 residue of Pur6 are indicated as yellow spheres.

Incorporation of an amino acid residue in the biosynthetic product is common in the biosynthesis of secondary metabolites, especially those encoded by nonribosomal peptide synthetase (NRPS) BGCs. However, classical amino acid activation mechanisms such as standalone adenylation domains or ATP-binding domains were not detected in BGCs of puromycin-like secondary metabolites. Therefore, no catalytic activity responsible for the activation of the amino acid precursor was detected in both Pur6 and puromycin-like BGCs. In this regard, a particularly important question about the mechanism of the amino acid-nucleotide coupling reaction arises. A detailed analysis of the BGC genomic context revealed cysteine tRNA encoded at the end of puromycin-like BGCs in *N. kunsanensis* and *X. bovienii* (Figure 2B). Similar structures were detected at the end of puromycin-like BGCs in *Photorhabdus temperate* and other *Xenorhabdus*, including *Xenorhabdus nematophila* (Figure 2C). Hence, trying to find an activated amino acid precursor, we hypothesize that aminoacyl-tRNA may serve as an amino acid donor in the coupling reaction catalyzed by Pur6-like enzymes (Figure 2D). Taking into account structural homology with cysteine proteinases, we suggest that Pur6 uses aminoacyl-tRNA as a substrate, and Cys326 is involved in the active intermediate formation that simulates the common mechanism of acyl activation by the acyl carrier protein (ACP) domain in NRPSs and PKSs. If this hypothesis is correct, the biosynthetic product of puromycin-like BGCs in *N. kunsanensis*, *X. bovienii*, and *S. monomycini* should contain cysteinyl residue.

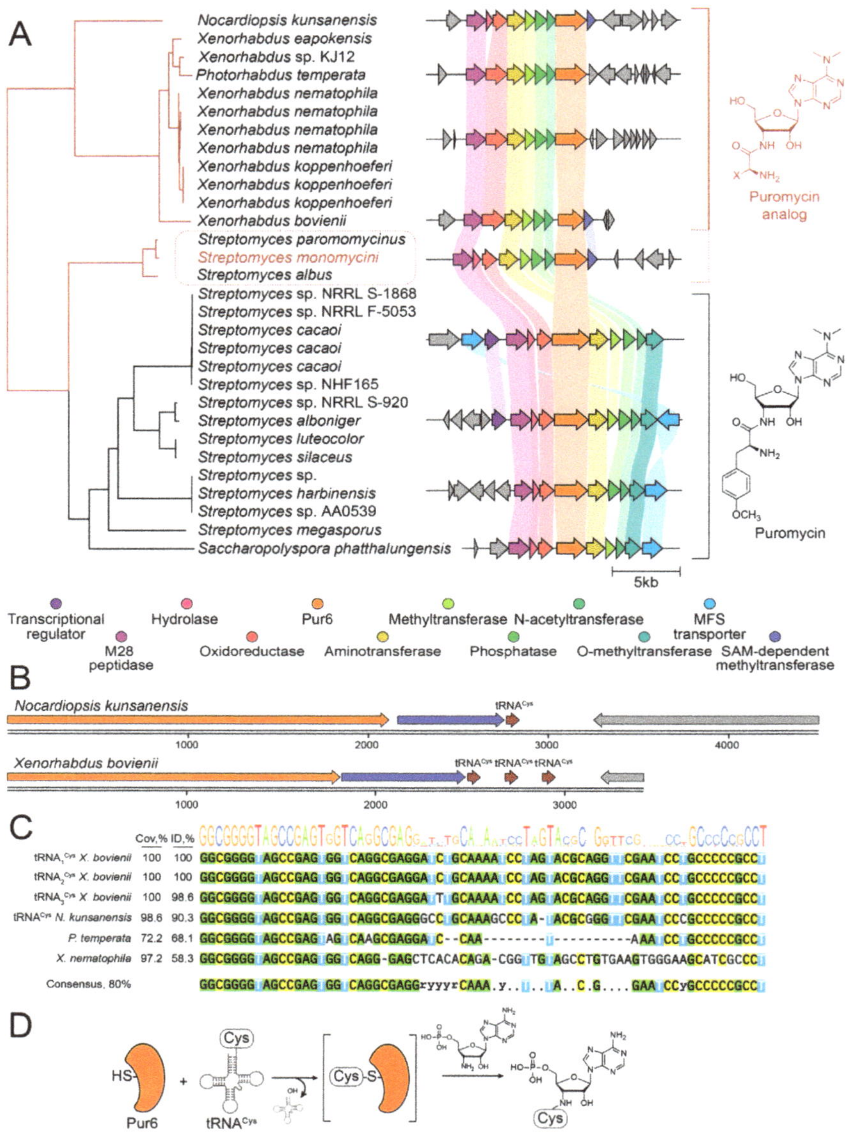

Figure 2. Pur6-guided genome mining revealed the biodiversity of puromycin analogs in *Streptomyces* and *Xenorhabdus*. (**A**) Pur6 and its homologs are strongly associated with BGC of puromycin-like secondary metabolites. The phylogenetic tree indicates the distance between Pur6 homologs in various bacteria. The corresponding genomic context and puromycin BGCs are colored with grey and rainbow, respectively. Putative BGCs associated with puromycin biosynthesis and biosynthesis of puromycin-like analogs are united by brackets and colored with black and red, respectively. Putative new analogs of puromycin in streptomycetes are highlighted with a red dotted line. (**B**) BGCs of puromycin-like secondary metabolites in *Nocardiopsis kunsanensis* and *Xenorhabdus bovienii* have cysteine tRNA. (**C**) Sequences of $tRNA^{Cys}$ and similar RNA sequences encoded in the genomic context of BGCs of puromycin-like secondary metabolites. (**D**) Putative reaction scheme illustrating a potential working principle of Pur6-like enzymes.

2.2. Cystocin Is a Biosynthetic Product of Puromycin-like BGC

To investigate the exact biosynthetic product of puromycin-like BGCs in the aforementioned streptomycetes, we addressed the All-Russian Collection of Microorganisms (VKM) for taxonomically related strains (*S. albus*, *S. monomycini* and related *Streptomyces* sp.). The strain *Streptomyces* sp. Ac-502 was positive for the production of antimicrobials. It was cultivated and subjected to solid-phase extraction and activity-guided fractionation. The active component was found to have UV maxima at 214 and 276 nm (Figure 3A). Mass spectrometry revealed an $[M+H]^+$ ion with an exact mass of 412.1767 Da, corresponding to the molecular formula $C_{16}H_{25}N_7O_4S$, which is consistent with a cysteinyl-containing analog of Puro. MS fragmentation (Figure 3C) in the positive mode HCD fragmentation showed predominant fragment ions at m/z 249.09 and 164.09 due to *N,N*-dimethyladenosine loss. The ion 118.03 was assigned as S-methyl-cysteine moiety. All product ions containing the primary amino group had a characteristic loss of a 17-amu fragment corresponding to the mass of ammonia. For further structure elucidation, the compound was analyzed with NMR.

The NMR data (Figure 3B, Supplement S1) confirmed that the isolated compound is the known antibiotic cystocin (Cst) that was previously identified in *Streptomyces* sp. GCA0001 [24].

2.3. Complete Genome Sequencing and Analysis of Cst BGC

The phylogenetic analysis using the 16S rRNA gene sequence derived from the genome assembly revealed that the Cst-producing strain *Streptomyces* sp. VKM Ac-502 was closely related to *S. monomycini* NRRL B-24309^T, *S. ochraceiscleroticus* NRRL ISP-5594^T, and *S. violens* NRRL ISP-5597^T (Figure S1). The analysis of the complete genome sequence showed the aminonucleoside BGC, which was named the *cst* BGC. Cst belongs to a diverse group of peptidyl nucleoside antibiotics [25]. The compounds of similar biosynthetic origin contain a nucleoside core and peptidyl moiety (Figure 4A). While the biosynthesis of Puro has been studied in detail [10–21], as far as we know, the biosynthesis of Cst has not been reported.

The overall architecture (Figure 4, Supplement S2) of the *cst* cluster and its protein similarity were analyzed and compared primarily with established peptidyl nucleosides, particularly with Puro (Table S1, Figure 4B).

Cst BGC contains a set of enzymes (CstH, CstG, CstF, CstE, CstD, CstC), homologs to those involved in the assembly of the nucleoside moiety of Puro in *pur* BGC (Figure 4B). The nucleoside moiety of a similar antibiotic, A201A, is also assembled by the set of homologous enzymes (ataP3/pur3, ataP4/pur4, ataP5/pur5, ataP7/pur7, and ataP10/pur10) [26–29]. Moreover, a recently isolated antibacterial prodrug mimic of GTP, ADG, possesses a nucleoside nature [30]. Its BGC includes homologs of the *pur7*, *pur10*, *pur3*, and *pur4* genes, which indicates their participation in the formation of the nucleoside core of the molecule.

The enzyme encoded by the *cstG* gene displays a 61% similarity to the product of *pur10*, which participates in the conversion of a hydroxyl group to a carbonyl group during puromycin biosynthesis [11]. A similar activity was observed for the product of the *rifL* gene during rifamycin biosynthesis, which has a high degree of similarity at the amino terminus (with identity 32%) with the product of *pur10* [31,32]. CstG, along with Pur10 [10], functions as an NAD-dependent ATP dehydrogenase.

CstH exhibits a 64% identity with the product of the *pur7* gene in the *pur* BGC. Pur7 is a nudix hydrolase [20] responsible for producing pyrophosphate and the relevant NMP from NTP. CstF is an aminotransferase that is likely involved in the subsequent formation of 3′-amino-3′-dAMP, showing a similarity of 78% to Pur4. The dephosphorylation of the nucleoside core of Puro is performed by the product of the *pur3* gene (with a similarity of 69% to CstD), which is essential for puromycin biosynthesis [21].

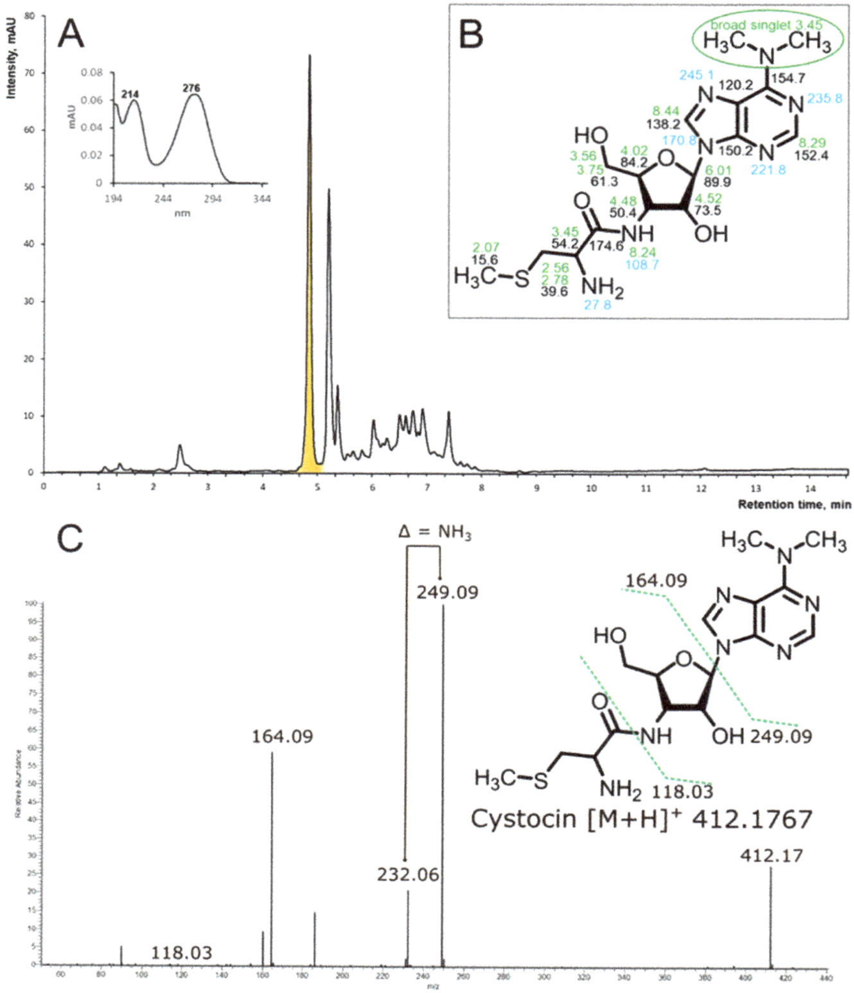

Figure 3. (**A**) rp-HPLC of the active fraction (Agilent HC-C18(2) 4.6 × 150 mm, 5 µm, eluent: solvent A—10 mM NH_4OAc, pH 5, solvent B—MeCN, gradient elution from 15 to 90 solvent B for 10 min, flow rate 1.5 mL/min, UV 275 nm, sample volume 2 uL, active component is highlighted with yellow. (**B**) The structure of Cst with the indication of NMR chemical shifts in DMSO-d6 at 35 °C. Green numbers stand for the ^{1}H chemical shifts, black—for the ^{13}C, and blue—for ^{15}N. Chemical shifts are provided in ppm, measured with respect to trimethylsylan as an external reference standard. ^{13}C and ^{15}N shifts were referenced indirectly, using the gyromagnetic ratios as implemented in Topspin software (Bruker Biospin Gmbh, Germany). (**C**) The positive mode HCD mass spectra of Cst and proposed fragmentation of the parent ion of the compound at m/z 412.17.

Cst and Puro have different types of methylation, S- and O-methylation, respectively. Hence, the *cst* BGC does not have a homolog of the *dmpM* gene, which is required for the O-methylation of tyrosine [17]. The proposed S-methylation of cystocin is mediated by the product of the *cstA* gene—a methyltransferase that exhibits low identity to the O-methyltransferase DmpM.

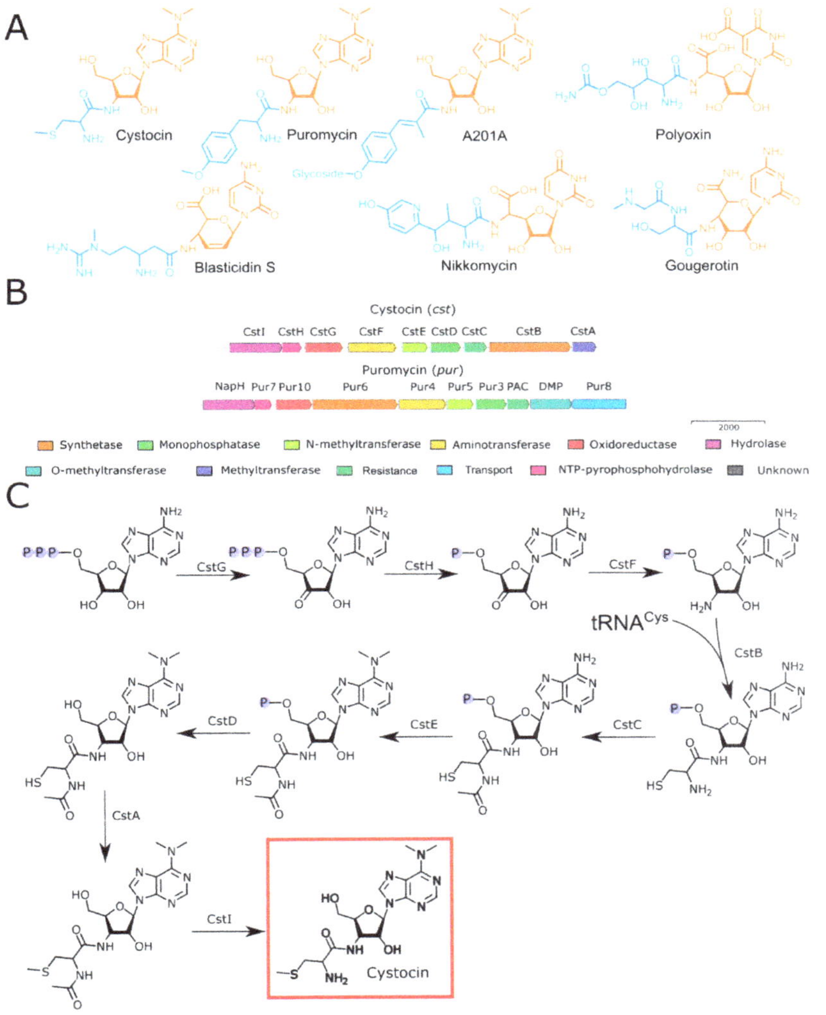

Figure 4. Biosynthesis of aminonucleosides. (**A**) Structures of Cst and related compounds. Nucleoside cores are highlighted in orange, while peptidyl fragments are marked in blue. (**B**) The suggested cst BGC (Orf1-Orf20) with some proposed functions and the comparison with *pur* BGC from *Streptomyces alboniger* strain ATCC 12461 sequence (NZ_CP023695.1). (**C**) The proposed biosynthetic pathway of Cst.

The attachment of the tyrosine moiety to the nucleoside core of Puro is mediated by Pur6 [11,12]. Within the *cst* BGC, the introduction of Cys is mediated by an aminonucleoside synthase referred to as CstB. CstB has modest similarity (53%) to Pur6, presumably also involving $tRNA^{Cys}$ as an amino acid substrate. This type of aminonucleoside synthase is specific for Puro-like antibiotics. tRNA-dependent peptide bond formation, mediated by the transferase PacB, has been observed in the biosynthesis of pentapeptidyl nucleoside antibiotics pacidamycins [33]. Enzymatic machinery responsible for assembling the nucleoside and peptidyl moieties is different from other peptidyl nucleoside antibiotics. Most of the aminonucleosides are synthesized with ANP-grasp-fold superfamily enzymes (NikS for nikkomycin [34], BlsL for blasticidin S [35], SanS and PolG for nikkomycin analogs [36] and polyoxins [37]). Members of this superfamily possess a unique structure in their ATP-binding sites and catalyze the ATP-dependent coupling of a carboxylic acid with a nucleophile, with the formation of an acylphosphate intermediate [38]. Alternatively, peptidyl moiety can be activated with adenylation (e.g., NpsA in biosynthesis of streptothricins [39]) or CoA-binding (e.g., GouK-activated seryl or sarcosyl groups are transferred by an acyl-CoA N-acyltransferase GouJ from acyl-CoA to the amino group of the gougerotin nucleoside moiety [40]).

Resistance to Puro is mediated by the N-acetyltransferase PAC [14]. Puro is initially produced as an inactive form (N-acetylpuromycin). This compound is actively transported out of the cell, and on the cell's exterior, the acetamide group is hydrolyzed by the N-acetylhydrolase NAPH, releasing the active Puro into the surrounding environment [15]. Within the *cst* BGC, we have identified an N-acetyltransferase CstC, which shares a significant similarity (66%) with N-acetyltransferase PAC. Hence, we suggest that CstC also inactivates Cst by N-acylation. In turn, CstI mediates the reactivation of Cst since it has a moderate similarity (54%) with NapH [11].

2.4. CstC Mediates Resistance to Cst

To understand the differences in functionality of aminonucleoside-inactivating N-acetyltransferases, PAC and CstC were studied. The CstC AlfaFold 3 structure model is very similar to PAC, having an RMSD = 0.53 Å. Two amino acid substitutions in the CstC model, A31F and L49F, were located at a distance less than 3 Å and 1 Å to the O-methyltyrosine residue of Puro, resulting in a more bulky active site and steric hindrance for Puro in CstC (Figure 5A,B). Steric hindrance of the antibiotic-binding pocket results in CstC preference for cystocin as a substrate.

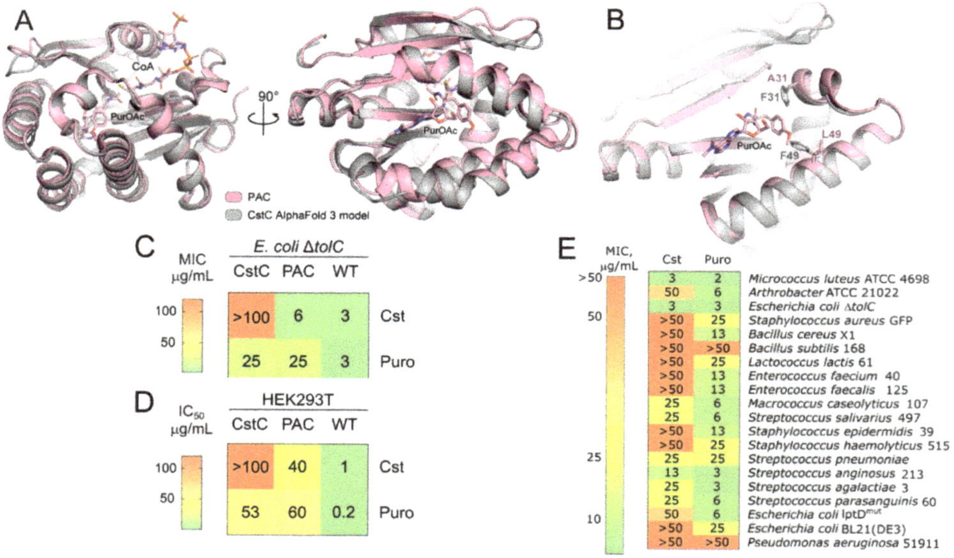

Figure 5. CstC mediates resistance to cystocin. (**A**) Structural alignment of the CstC AlfaFold 3 model (grey) with the PAC-PuroOAc-CoA complex (pink), PDB ID 7K0A. (**B**) Antibiotic-binding pocket of CstC and PAC. The 31 and 49 positions located proximately to the aminoacyl residue of Puro-OAc are highlighted with sticks. (**C**) Comparison of bacterial susceptibility for *E. coli* Δ*tolC* and *E. coli* Δ*tolC* producing CstC or PAC. (**D**) Cytotoxicity of Cst and Puro toward human cell line HEK293T and HEK293T producing CstC or PAC. (**E**) Heat map of MICs of Cst and Puro against Gram-positive and Gram-negative bacteria.

Aminonucleosides have potent antibacterial activity against hypersensitive bacterial strains (Figure 5C) and cytotoxicity toward mammalian cells (Figure 5D, Table S2). Recombinant production of CstC in *E. coli* Δ*tolC* mediates its resistance to Cst, increasing MIC values more than 16 folds (Figure 5C). It also provides partial resistance toward Puro. On the contrary, recombinant production of PAC does not influence *E. coli* Δ*tolC* resistance to Cst, making the bacteria resistant to Puro (Figure 5C). Even higher resistance of N-acetyltransferase-producing mammalian cells was observed (Figure 5D). PAC-producing cells were resistant to both Cst and Puro, although the acquired resistances were higher for Puro than Cst. CstC-producing cells, in their turn, acquired higher resistance to Cst (Figure 5C,D).

To assess natural bacterial resistance to Cst and Puro, susceptibility testing against a wide range of Gram-positive and Gram-negative bacteria, including hypersensitive strains of *Escherichia coli* $lptD^{mut}$ and Δ*tolC*, was performed (Figure 5E). Cst is active against *Micrococcus luteus* ATCC 4698, *E. coli* $lptD^{mut}$, *E. coli* Δ*tolC*, *Arthrobacter* ATCC 21022, and *Macrococcus caseolyticus* 107. The activity of Puro was 2–8 times higher than that of Cst, and it has a broader activity spectrum. A high activity against *E. coli* Δ*tolC* compared with *E. coli* $lptD^{mut}$ suggests that Cst resistance is mediated by efflux in Gram-negative bacteria.

2.5. Cst Acts by Translation Inhibition

Puro inhibits translation by incorporation into the growing peptide chain, leading to the premature release of puromycylated peptides (Figure 6A). The Cst antibacterial mechanism was first assessed using reporter *E. coli* strain JW5503 Δ*tolC* pDualrep2, which consists of two fluorescent protein reporter genes, *turborfp* (induced by SOS-inducible *sulA* gene promoter) and *katushka2S* (regulated by the modified *trpL* attenuator sequence) [41]. Expression of Katushka2S under treatment with a sublethal concentration of Cst suggested protein synthesis inhibition by the isolated compound (Figure S2).

Cst efficiently inhibited protein synthesis in both prokaryotic and eukaryotic cell-free translational systems (Figure 6B,C). Cst has a comparable activity level with Puro in blocking bacterial protein synthesis (0.9 ± 0.4 μM and 0.2 ± 0.1 μM for Cst and Puro, respectively), being 8-fold less active in the eukaryotic system (3 ± 2 μM and 0.4 ± 0.4 μM for Cst and Puro, respectively).

To understand the mechanical details of the inhibition of the peptidyl-transfer reaction, visualization of the synthesized fluorescently labeled peptides is provided (Figure 6D). In line with the previous results, Puro was more active than Cst, inhibiting the synthesis of very small MFF peptides (Supplement S3). For longer MFFFF peptides, Cst inhibition was more efficient, indicating more sufficient synthesis blocking of long templates (Figure 6E). The formation of cystocinilated and puromycinilated truncated peptides was corroborated with LCMS analysis (Supplement S4).

The toe-printing analysis confirmed similar modes of translation inhibition by Cst and Puro (Figure 6F). In high concentrations (Figure 6G), both Puro and Cst inhibited translation at the start codon or near it. However, in the case of Puro, the inhibition occurs completely so that ribosomes do not reach the threonine codon, while for reaction with cystocin, a part of ribosomes reaches the threonine codon and we can detect ribosome capture at the Thr codon caused by borrelidin. Therefore, Cst is more important in blocking the synthesis of large proteins.

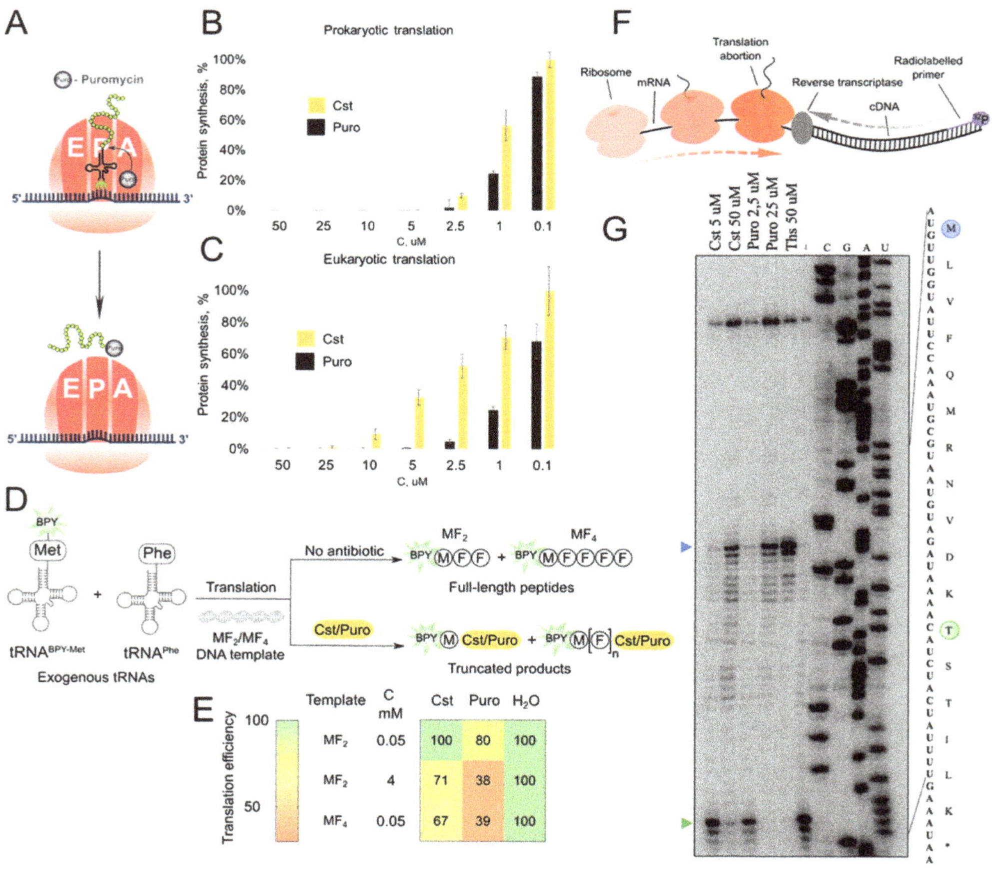

Figure 6. Cst mode of action. (**A**) Schematic representation of Puro mode of action. In vitro prokaryotic (**B**) and eukaryotic (**C**) protein synthesis inhibition by Puro and Cst. (**D**) Fluorescently labeled short peptides synthesis inhibition assay scheme, labeled peptides MF_2 and MF_4 are synthesized with coupled transcription–translation from exogenically formed loaded tRNA and visualized using PAGE with fluorescent detection. BPY—BODIPY. (**E**) Heatmap of normalized intensity of full-length peptides under treatment with Cst/Puro. (**F**) The principal scheme of toe-printing analysis. (**G**) Toe-printing analysis for Cst/Puro. The green arrow indicates the position of ribosome capture on the Thr codon; the blue arrow indicates the start codon. Ths—thiostrepton.

3. Discussion

Despite the widespread use of puromycin in biotechnology, its biosynthesis is still not described in molecular details. Genome mining based on the key biosynthetic enzyme—unique aminonucleoside synthase Pur6—provides clues for its activity. We propose that Pur6-like enzymes use aminoacyl-tRNAs as a substrate for the amino acid-nucleotide coupling reaction. In this case, *Streptomyces* may utilize their genome-encoded tRNAs. In contrast, *Xenorhabdus* appears to have acquired puromycin BGC through horizontal gene transfer from *Streptomyces* or *Actinomycetia*, such as *N. kunsanensis*, and adapted it by incorporating additional tRNA, as tRNAs in Gram-positive and Gram-negative bacteria are not interchangeable for Pur6-like enzymes.

Analysis of related clusters allowed us to evaluate and access the natural structural diversity of aminonucleoside antibiotics, revealing the *cst* BGC. Distinct BGC architecture and its association with $tRNA^{Cys}$ allowed us to cluster Cst-producing bacteria, including *Streptomyces* sp. VKM Ac-502. In silico analysis of the BGC and structural elucidation of the biosynthetic product both confirmed the production of the cysteinyl-containing Puro analog, i.e., Cst. Notably, there exists a naturally occurring bacterial puromycin-related metabolite with a 3′-N-amino acid substitution differing from the classical 3′-N-tyrosinyl. In the case of puromycin B, a leucine residue replaces the tyrosine. The biosynthesis of such compounds involves a promiscuous aminonucleoside synthetase [42]. Further exploration of the amino-nucleoside synthase could potentially lead to the discovery of novel peptidyl nucleoside antibiotic congeners. Our findings suggest that the directed evolution of Pur6-like enzymes could expand the chemical space of the puromycin antibiotic family by diversifying the peptidyl moieties.

The biosynthesis of aminonucleoside antibiotics includes specific self-resistance mechanisms, such as N-acetylation of the produced antibiotic, rendering it inactive within the producing organism. Puro-based systems for the selection of cell lines utilize this resistance mechanism. The development of modified variants of PAC for improved selection systems still attracts significant attention [43]. The *cst* BGC contains an N-acetyltransferase, CstC, with a similar function. The specificity of CstC is different from that of the PAC enzyme. Susceptibility testing revealed that both CstC and PAC provide some level of cross-resistance. However, the acquired resistance is significantly higher for the corresponding aminonucleoside. This effect is mediated by the bulky substituents in close proximity to the catalytic center of CstC, making it less spatially accessible for Puro. This finding could provide a basis for the rational design of selective resistance enzymes to develop selection systems orthogonal to puromycin-based ones.

Puro is a well-known translation inhibitor and is widely used in scientific research. Cst is similar to Puro in many ways: it is toxic to Gram-positive bacteria, suppresses protein biosynthesis in vitro in cell-free translation systems, and at high concentrations, even inhibits the biosynthesis of short peptides. Despite structural homology to tyrosyl-tRNA, Puro incorporation is known to be not amino acid specific [5]; the same mode of action was observed for Cst. Analyzing the toe-prints and small peptides labeled with BODIPY, we came to the conclusion that the effect of both puromycin and cystocin is cumulative. That is why the longer the template, the more frequently the transfer occurs. On short templates, the inhibition effect is noticeable only at very high concentrations, when aminonucleoside saturates all ribosomes. The obtained results generalize the diversity of aminonucleoside synthetases, providing a basis for the further discovery of new aminonucleosides. Moreover, the discovered mechanistic details could be applied to expand the biosynthetic potential of Pur6-like enzymes and provide selective aminonucleoside-inactivating enzymes.

The findings presented in this manuscript highlight the significant biotechnological potential of aminonucleoside-based systems for further development in molecular biology. However, the currently obtained results come with certain limitations, which hinder their immediate practical application. The first studied enzyme, aminonucleoside synthase, demonstrates the potential for directed evolution to produce novel aminonucleoside antibiotics with unique biological functions. Additionally, this study successfully obtained only one strain from the cst-like BGC-bearing clade, suggesting that further mining of the strains related to the identified *Streptomyces* and *Xenorhabdus* species could uncover new natural congeners with potentially distinct properties.

Moreover, the self-resistance enzyme CstC shows promise as a candidate for the development of an orthogonal selective marker in molecular biology. However, the current Cst/Puro selectivity parameters require further optimization. The identified amino acid substitutions, A31F and L49F, which introduce steric hindrance in the active site, represent promising starting points for the engineering of highly selective N-acetyltransferases. Future studies focusing on enzyme optimization and exploring the diversity of aminonucleoside-producing strains could unlock new avenues for the practical application of these systems in biotechnology.

4. Materials and Methods

4.1. Genome Sequencing and Data Analysis

Pur6 homologs were identified among non-redundant protein sequences using protein BLAST [44]. Pur6-like enzymes had E-value > 1×10^{-40}, cover > 76%, protein identity > 28%, similarity > 40%. Pur6-like enzymes were downloaded from GenBank with adjacent 100 kb genome fragments and processed with antiSMASH to identify BGCs [45]. A tree of Pur6-like enzymes was built with MAFFT [46]. BGC analysis was performed with CAGECAT [47]. tRNAs were detected with tRNAscan-SE 2.0 [48], and alignments were visualized with MView [49]. Structure models were built with AlphaFold 3 [50]. Protein sequences with the following accession numbers were used for phylogenetic tree construction: WP_026116056.1, WP_074025210.1, WP_021324692.1, WP_099109202.1, WP_010848540.1, WP_041976841.1, WP_013185625.1, CEE90880.1, WP_339351399.1, SFU95741.1, WP_245759091.1, WP_258087068.1, WP_030021787.1, WP_125058165.1, WP_030888603.1, WP_030885422.1, WP_301126629.1, WP_086817987.1, WP_149564235.1, WP_159789347.1, WP_030778222.1, WP_055528327.1, WP_069885355.1, WP_055700141.1, WP_093842894.1, WP_176127489.1, WP_019433835.1, WP_031508068.1, WP_184730161.1.

The DNA was isolated using the Wizard DNA extraction kit (Promega Corporation, Madison, WI, USA) and size-selected with optimized solid-phase reversible immobilization (SPRI) beads. The DNA concentration and quality were determined on a Qubit 4 Fluorometer and Nanodrop ND-1000 (Thermo Fisher Scientific, Waltham, MA, USA). The long reads were generated with MinION sequencing (Oxford Nanopore Technologies, Oxford, UK). The sequencing libraries were prepared using the ligation sequencing kit SQK-LSK109 and native barcoding expansion kit EXP-NBD114 and run in a FLO-MIN106 flow cell. Reads were basecalled using Guppy v3.6.1. with default parameters (guppy_basecaller). A NEBNext Ultra DNA library prep kit (New England Biolabs, Ipswich, MA, USA) was used to prepare fragment libraries for genome sequencing. Sequencing was performed on the HiSeq 2500 system (Illumina, San Diego, CA, USA) HiSeq SBS Kit v4 (250 Cycle) using a 2 × 125 bp run configuration. De novo assembly was performed by Flye assembler (2.8.1) [51] using default parameters. Illumina reads were used to improve and correct genome assembly using the Pilon program [52]. Identification of the protein-coding sequences and primary annotation were performed using PROKKA v1.14.6 [53]. The draft genome sequences and raw sequencing reads have been deposited at GenBank, BioProject accession number PRJNA1158227, GeneBank accession number JBHFFO000000000.

The genome of *Streptomyces* sp. VKM Ac-502 was analyzed with antiSMASH to identify potential BGCs [45]. Homologous gene clusters were identified with MultiGeneBlast [54] using the MIBiG database [55]. Putative functions of enzymes were assigned according to the predicted function of the closest characterized relative identified by Blast search [44] in NCBI.

A truncated puromycin BGC sequence was acquired from the MiBIG database (BGC0000878). Complete puromycin BGC was acquired from *Streptomyces alboniger* ATCC 12461 sequence (NZ_CP023695.1). Additional tailoring enzyme sequences were acquired from Genbank: *pur8* (X76855.1), *dmpM* (M74560.2), and *pac* (M25346.1).

The generic affiliation of *Streptomyces* sp. VKM Ac-502 was initially determined on the basis of a 16S rRNA sequence derived from the genome assembly of strain. The 16S rRNA gene sequence (1451 bp) similarities between the target strain and the type strains of known *Streptomyces* species were estimated using the EzBioCloud server (https://www.ezbiocloud.net (accessed on 20 August 2024) [56]). A digital DNA-DNA hybridization (dDDH) [57] and Average Nucleotide Identity (ANI) values [58–60] between VKM Ac-502 and type strains of *Streptomyces* species were calculated using the Type (Strain) Genome Server (https://tygs.dsmz.de/user_requests/new (accessed on 20 August 2024) [61]) and JSpecies WS tool (https://jspecies.ribohost.com/jspeciesws (accessed on 20 August 2024) [62]), respectively.

4.2. Bacterial Strains and Cell Lines

The antibacterial activity of the strains was evaluated using a hypersensitive *E. coli* Δ*tolC* strain transformed with pDualrep2 reporter plasmid [41]. The cystocin-producing strain *Streptomyces* sp. VKM Ac-502, also known under the invalid species name "*Streptomyces tumemacerans*" [63,64], was obtained from the All-Russian Collection of Microorganisms (www.vkm.ru).

A bacterial collection including *Micrococcus luteus* ATCC 4698, *Bacillus cereus* X1, *Lactococcus lactis* 61, *Enterococcus faecium* 40, *Enterococcus faecalis* 125, *Macrococcus caseolyticus* 107, *Staphylococcus epidermidis* 39, *Staphylococcus haemolyticus* 515, *Pseudomonas aeruginosa* 51911, *Streptococcus pneumonia* ATCC 6303, *Streptococcus anginosus* 213, *Streptococcus agalactiae* 3, *Streptococcus parasanguinis* 60, and *Streptococcus salivarius* 497 was kindly provided by Lytech Co., Ltd. (Moscow, Russia) [65]. Staphylococcus aureus GFP constitutively producing GFP was kindly provided by Andrey Shkoporov from the Department of Microbiology and Virology, Russian National Research Medical University, Moscow. The *E. coli* ΔtolC KanS, *E. coli* lptD^{mut}, and *E. coli* strain JW5503 (ΔtolC) [66] transformed with pDualrep2 reporter plasmid [41] were used from our laboratory strain collection. Strains *Arthrobacter* sp. ATCC 21022, *Bacillus subtilis* 168, and *Escherichia coli* BL21(DE3) were purchased from Invitrogen Corp. (Carlsbad, CA, USA). Cell line HEK293T cells were kindly provided by Dr. E. Knyazhanskaya.

4.3. Cultivation, Isolation, and Purification of Cystocin

The strain *Streptomyces* sp. VKM Ac-502 was transferred from the surface of the agar medium into a 750 mL Erlenmeyer flask with 50 mL of LB nutrient medium of the following composition (components content in g/L): tryptone—10, yeast extract—5, NaCl—10, distilled water—up to 1 L, pH 7.4. Cultivation was carried out at 28 °C for 4 days on a thermostat shaker Innova 40 (New Brunswick Scientific, Middleton, WI, USA) at 150 rpm. The seed culture of the second generation was grown in 150 mL of LB medium under similar conditions for 6 days, using the first inoculum in an amount of 3% vol. The strain produced cystocin starting from 2 days of cultivation and reached its maximum production on day 6 of cultivation.

Bacterial cells were eliminated from culture broth by centrifugation at 5000 rpm on a Sigma 3–16KL centrifuge and filtration through a 0.47 μm MCE membrane filter (Millipore, Billerica, MA, USA). A total of 1 L of clarified supernatant was loaded on the 30 mL cartridge packed with 7 g of LPS-500-H polymer sorbent (copolymer divinylbenzene—hydrophilic monomer, pore size 50–1000 Å, 70 μm, Technosorbent, RF) at flow rate 15 mL/min using a peristaltic pump (Masterflex L/S variable speed pump systems, Masterflex, Gelsenkirchen, Germany). Extraction by a water/acetonitrile (MeCN) mixture was performed at a flow rate of 15 mL/min in gradient mode from 0 to 100% ACN for 10 min using a PuriFlash 5.250 instrument (Interchim, Montluçon, France), and 15 mL fractions were collected. The activity of the collected fractions was tested using the reporter *E. coli* Δ*tolC* pDualrep2. The most active fraction (eluted at about 20% MeCN) was further analyzed using HPLC on an RP column using a Nexera X2 LC 30A instrument (Shimadzu, Kyoto, Japan) equipped with an SPD-M20A detector (Shimadzu, Kyoto, Japan). HPLC conditions were as follows: column Zorbax SB C18 4.6 × 150 mm, 5 μm, eluent solvent A—10 mM NH_4OAc, pH 4.5, solvent B—MeCN; gradient elution from 5 to 90% of solvent B; flow rate 1.5 mL/min; UV detection at 275 nm. HPLC fractions were collected and tested for activity, and the fraction containing pure active substance (Figure S3) was isolated and then analyzed by LCMS. Isolation of the active compound for NMR experiments and biological assays was performed using semipreparative column Zorbax SB C18 9.4 × 150 mm, 5 μm, with solvents A and B as above at 16% of solvent B in isocratic mode and at the flow rate 2.5 mL/min.

4.4. LCMS Analysis

LC-MS analysis was carried out on an Ultimate 3000 RSLCnano HPLC system connected to an Orbitrap Fusion Lumos mass spectrometer (ThermoFisher Scientific, Waltham, MA, USA) with the loading pump used for analytical flow gradient delivery. Samples were separated on a Gemini NX-C18 3 μm 100 Å column 100*2.1 mm at 200 μL/min flow rate in the linear gradient of acetonitrile in water with the addition of 10 mM ammonium formate and 0.1% FA. UV data were collected at 275 nm. MS1 and MS2 spectra were recorded at 30 K and 15 K resolution, respectively, with HCD fragmentation. Raw data were collected and processed on Thermo Xcalibur Qual ver. 4.3.73.11.

4.5. Structure Elucidation

The structure of cystocin was elucidated using the conventional heteronuclear NMR approach. A total of 1.5 mg of pure compound was dissolved in 350 uL DMSO-d_6 (99.98% purity, CIL, Tewksbury, MA, USA) and placed into the Shigemi DMS-005TB NMR tube (Shigemi, Kawaguchi, Japan). NMR spectra were recorded at 35 and 45 °C using the Bruker AvanceIII 800 MHz NMR spectrometer (Bruker, Billerica, MA, USA) equipped with a TCI cryogenic probe. The 1D ^{1}H,^{13}C and 2D ^{1}H,^{13}C-HSQC; ^{1}H,^{15}N-HSQC; ^{1}H,^{13}C-HMBC (optimized for 8 Hz J-coupling); ^{1}H,^{15}N-HMBC (optimized for the 5 Hz J-coupling), DQF-COSY, ^{1}H,^{13}C-HSQC-TOCSY (20 ms mixing) and ROESY (200 ms mixing) spectra were recorded. The physico-chemical properties of the compound were identical to previously reported [24].

4.6. Plate Test Against Escherichia coli Reporter Cells

The *E. coli* strains (JW5503 Δ*tolC* and *lptD^{mut}*) transformed with pDualrep2 reporter plasmid were used to evaluate the mechanism of antimicrobial action and activity against bacteria. The overnight culture of the reporter strains was diluted with fresh LB medium to an optical density of 600 nm (OD600) of 0.05–0.1. The cultures were transferred to LB agar plates with 100 μg/mL ampicillin or 50 μg/mL kanamycin applied. On an agar plate with the lawn of the reporter strains, 1 μL of compounds Cystocin (Cst) and Puromycin (Puro) at a concentration of 20 mg/mL were spotted. Erythromycin (Ery, 5 mg/mL) and Ciprofloxacin (Cip, 10 μg/mL) were used as control antibiotics on an agar plate. ChemiDoc (Bio-Rad, Hercules, CA, USA) was used to scan agar plates in the Cy3 (TurboRFP) and Cy5 (Katushka2S) channels after overnight incubation at 37 °C. This was performed to determine the inhibition zone and fluorescence levels of the reporter proteins.

4.7. Antibacterial Activity Assessment

Determination of the activity spectra of Cst and Puro for comparative analysis was carried out using the method of serial dilutions. Individual colonies of bacteria from the plate with 2YT agar medium (10 g/L yeast extract, 16 g/L tryptone, 5 g/L NaCl, 15 g/L agar) were transferred to 5 mL of 2YT nutrient medium (10 g/L yeast extract, 16 g/L tryptone, 5 g/L NaCl) and incubated at 37 °C overnight. The resulting overnight cultures were transferred to fresh nutrient medium in a ratio of 1:100 and incubated at 37 °C for 3 h. Next, the cultures were diluted so that the optical density of the final solution at 600 nm was ~0.001 a.u. This cell culture was used to study the spectrum of antibiotic activity (Cst and Puro). Antibiotic MICs were determined after 16 h of incubation in a 96-well plate at a wavelength of 600 nm using a Varioskan multimodal reader (Thermo Fisher Scientific, USA); 50 μg/mL was chosen as the initial concentration for antibiotics with subsequent 2X-fold serial dilutions.

4.8. Cytotoxicity Studies

To prepare the MTT solution, a dry sample of thiazolyl blue tetrazolium bromide was dissolved in phosphate-buffered saline, pH = 7.4 (DPBS), to a concentration of 5 mg/mL. The resulting solution was sterilely filtered through a filter with a pore diameter of 0.2 μm, and the filtrate was collected in a sterile container protected from light. To prepare a solubilizing solution, 40% (vol.) dimethylformamide (DMF) was diluted in 2% (vol.) glacial acetic acid. A total of 16% (*w*/*v*) sodium dodecyl sulfate (SDS) was added to the resulting solution. Using 2% (vol.) glacial acetic acid, the pH of the solution was adjusted to 4.7.

Double dilutions of the test compounds were prepared in DMSO as a negative control in the culture medium. A final concentration series of eight 2-fold dilutions was started at 6.4 μg/mL. Incubation was carried out in a humidified atmosphere at 37 °C with 8% CO_2 for 72 h. Cytotoxicity was assessed in HEK293T cells [67]. Cell viability was determined by colorimetric assessment of cell metabolic activity (MTT). The analysis was carried out after incubation for 72 h. All measurements were performed in triplicate biological replicates. A total of 10 μL per well of MTT solution was added to the test samples to a final concentration of 0.45 mg/mL and incubated for an hour at 37 °C. Then, 100 μL of solubilizing solution was added to each well to dissolve the formazan crystals. The result was measured using a Varioskan multimodal reader (Thermo Scientific, USA) at an absorbance of 570 nm. The obtained results were analyzed using MS Excel 2013 software. Based on the results of the MTT test, the graph of the dependence of the average survival value on the concentration of the test substance plotted on a logarithmic scale was constructed. A well with cells that did not contain the test compounds was taken as 100% viability. Relative survival was calculated using the following formula: (absorption for a given well)/(absorption for a well without adding the drug)·100%. The cytotoxic concentration for each compound (IC_{50}) was determined from the resulting inverse sigmoid as the value of the concentration at which the cytotoxic effect is induced in 50% of the cells in the monolayer. Doxorubicin was used as positive control.

4.9. Resistance Studies

Construction of plasmids pSol_CstC and pSolPAC. The *Streptomyces* sp. VKM Ac-502 genomic DNA was used as a template for obtaining *cstC* sequence flanked with restriction sites for cloning. Primers for PCR were used as F_NdeI_CstC 5′ GCCAATCATATGAC-CTCGAACGCCCCCG 3′ and R_NotI_CstC 5′ TGACTGCGGCCGCTCAGGCGCCCGGCT-TACGG 3′.

The gene of puromycin N-acetyltransferase was amplified from plasmid pCDF1-MCS2-EF1-Puro (System Biosciences, Mountain View, CA, USA) with primers F_NdeI_PAC 5′ GCCtcaCATATGACCGAGTACAAGCCCACG 3′ and R_NotI_PAC 5′ TGACTGCGGC-CGCTCAGGCACCGGGCTTGCG 3′ flanked with restriction sites for cloning.

PCR products and vector pSolSumo (Lucigen, Middleton, WI, USA) were digested with restriction enzymes NdeI and NotI (New England Biolabs, Cambridge, UK), followed by ligation with T4 DNA ligase (Evrogen, Moscow, Russia) according to the manufacturer's protocol. The resulting plasmids pSol_CstC and pSol_PAC were transformed into *E. coli* DH10β and verified by sequencing. The *E. coli* Δ*tolC* kanS cells were transformed with the corresponding plasmids, and the obtained strains were used in antibiotic susceptibility experiments. In pSolSumo, vector gene expression is under the control of rhamnose-dependent rhaB promotor; therefore, different concentrations of rhamnose were used to control the CstC expression rate. However, no significant differences were observed.

Transformant antibiotic susceptibility testing. The minimal inhibitory concentrations (MIC) were estimated with a microwell broth dilution assay. In brief, the tested bacteria were diluted in 2YT media (10 g/L yeast extract, 16 g/L tryptone, 5 g/L NaCl) to a final concentration of approximately 1×10^6 CFU/mL in a 96-microwell plate, untransformed *E. coli* Δ*tolC* kanS were used as control. The cystocin and puromycin samples were used to form 2-fold serial dilutions starting with 100 μg/mL. Plates were incubated at 37 °C with shaking at 500 rpm for 16 h. After incubation, the optical density at 600 nm of each well was measured with a Varioskan Flash multimode reader (Thermo Fisher Scientific, USA). MICs were determined as the lowest antibiotic concentration that inhibited the growth of the bacteria. For expression induction, several concentrations of rhamnose were used (from 0 to 1 mM).

Mammalian cell growth and *pac* and *cstC* resistance genes transfection. The HEK293T cell line was cultured in a 6-well plate at 37 °C in 5% CO_2 in DMEM/F12 medium supplemented with 10% fetal bovine serum and Gibco GlutaMAX. Then, two pSBbi-NEO (AddGene 60525) vectors with the corresponding resistance genes were cloned by SfiI restriction followed by sticky ends insertion of the resistance genes obtained with PCR. The puromycin resistance gene, puromycin N-acetyltransferase (*pac*), was amplified from the Human CRISPR KO Library in lentiGuide-Puro (AddGene 1000000049). The cystocin resistance gene, *cstC*, was amplified from the genomic DNA of the producing strain *Streptomyces* sp. VKM Ac-502. Obtained vectors and transposase caring plasmid pCMV (CAT) T7-SB100 (AddGene 34879) were transfected into HEK293 cells using Lipofectamine 3000 (#L3000008, Thermo Fisher Scientific, USA) in accordance with the manufacturer's protocols, harvested at 24 h and treated with antibiotics at a concentration 5 times higher than the IC_{50}. Cell selection with incorporated resistance genes lasted 12 days, then cryostocks were prepared. Afterward, they were applied for further experiments.

4.10. In Vitro Studies

Bacterial in vitro translation assay. The PURExpress system (NEB) was used to conduct prokaryotic in vitro translation reactions. Each reaction was 5 μL in total, supplied with 0.1 mM of d-luciferin (Promega), 0.5 μL of either antibiotic solution or solvent (water), as indicated, and 100 ng of Fluc mRNA (the latter was added in 1 μL water solution after the reaction mixture supplemented with antibiotic was pre-incubated for 5 min at room temperature). After the mRNA addition, mixtures were transferred into a preheated white FB/NB 384-well plate (Grenier no. 781904) and incubated in the VictorX5 multi-reader at 30 °C with continuous measurement of luciferase activity.

Mammalian cell-free system. For testing compounds in a mammalian in vitro translation system, whole homemade HEK293T cell extracts were used. The total volume of the translation reaction was 10 μL, including 5 μL HEK293T extract, 1 μL 10× translation buffer (20 mM Hepes-KOH pH 7.6, 1 mM DTT, 0.5 mM spermidine-HCl, 0.8 mM $Mg(OAc)_2$, 8 mM creatine phosphate, 1 mM ATP, 0.2 mM GTP, 120 mM KOAc, and 25 μM of each amino acid), 2U of RiboLock RNase inhibitor (Thermo Scientific, USA), 0.5 mM d-luciferin (Promega), 1 μL of either antibiotic solution or solvent (water), as indicated, and 50 ng mRNA (the latter was added in 1 μL water solution after the reaction mixture supplemented with antibiotic was pre-incubated for 5 min at 30 °C). After adding the mRNA, the mixtures were transferred to a preheated white FB/NB 384-well plate (Grenier no. 781904) and incubated in the VictorX5 multi-reader at 37 °C with continuous measurement of luciferase activity.

Fluorescently labeled short peptides [68]. DNA templates containing short open reading frames coding M, MF, MF2, and MF4 short peptides were synthesized by PCR using the plasmid puC19 with corresponding inserts as a template for amplification. Each template harbored the T7 promoter and the SD sequence.

Coupled transcription–translation was set up in 10 μL reactions using a PURExpress Δ (aa, tRNA) Kit (NEB), with the addition of 20 ng of DNA template (M, MF, MF_2, and MF_4). Also, 0.1 μM BPY-Met-$tRNA^{fMet}$, 0.45 μM fMet-$tRNA^{fMet}$, 1 μM Phe-$tRNA^{Phe}$, 100 μM Phe, and 1 μL of either antibiotic solution or solvent (water), as indicated, were added to each reaction. The reactions were incubated for 30 min at 37 °C and divided into two tubes. To one part, 2 μL of 1 M $NaHCO_3$ was added, subsequently followed by incubation at 37 °C for 20 min. Reactions were stopped by adding formamide dye (95% formamide, 0.025% [w/v] bromophenol blue, 0.5 mM EDTA). Samples were then preheated for 3 min at 80 °C and loaded to a 10% denaturing PAGE (19:1 AA:bisAA; 1× TBE buffer; 8M urea). Gels were scanned by a Typhoon FLA 9500 Biomolecular Imager (GE Healthcare, Chicago, IL, USA) in the FAM channel with an excitation peak (493 nm) and emission peak (517 nm).

Toe-print analysis. The toe-printing assay was conducted according to the protocol described by Orelle et al. [69]. At the first stage, the primers were labeled with [γ-32P] ATP polynucleotide kinase (ThermoFisher, USA) according to the manufacturer's protocol. Next, in vitro translation of the short-model mRNA was performed using a PURExpress®

In VitroProtein Synthesis Kit (New England Biolabs, USA). The reaction mixture (volume, 5 μL) contained 2 μL of solution A, 1 μL of solution B, 0.2 μL of RiboLock (ThermoFisher, USA), 0.5 μL of the test compound, 0.5 μL of DNA template (0.2 mmol/μL), and 0.5 μL of the radiolabeled primer. A total of 50 μM of borrelidin was added to each sample to stop the translation on the threonine codon.

The mixture was incubated at 37 °C for 20 min, and 1 μL of the reverse transcription mix from the Titan One Tube RT-PCR System kit (Roche, Basel, Switzerland) was added. Reverse transcription was conducted for 15 min at 37 °C. The reaction was stopped by adding 1 μL of 10 M NaOH, followed by incubation at 37 °C for 15 min. The neutralization was performed by adding 1 μL of 10 N HCl. Next, 200 μL of the resuspension buffer was added. The resulting samples were purified using a QIAquick PCR purification kit (Qiagen, Hilden, Germany). The sequence mixtures were prepared using a USB® Thermo Sequenase Cycle Sequencing Kit (Affymetrix, Santa Clara, CA, USA) according to the manufacturer's protocol. Electrophoresis was carried out in 6% polyacrylamide gel (60 × 40 × 0.03 cm) containing 19% acrylamide, 1% N,N′-methylenebisacrylamide, and 7 M urea in TBE buffer for 2–3 h.

The specimens and products of the sequencing reactions (2 and 1.5 μL, respectively) were applied to the gel. The gel was transferred onto 3-mm paper, dried, and exposed to a sensory screen for 18 h. The screen was scanned using a Typhoon FLA 9500 Biomolecular Imager (GE Healthcare, Chicago, IL, USA). The ermCL template for this experiment was obtained by PCR amplification using a Taq-DNA-polymerase kit (ThermoFisher, USA), according to the standard protocol. The template sequence is described in Table S3.

Supplementary Materials: The following supporting information can be downloaded at https://www.mdpi.com/article/10.3390/ijms252312901/s1.

Author Contributions: Conceptualization, V.A.A., S.S.T., D.A.L. and P.V.S.; methodology, S.I.K., D.A.L., V.A.A., S.S.T. and O.A.B.; investigation P.A.Z., V.A.A., A.A.B., E.B.G., O.A.B., S.O.P., A.M.K., S.E.L., Y.A.P., E.A.M., V.I.M., E.A.N., E.S.K., I.P.S., O.V.B., L.I.E., E.V.A., K.S.M., V.V.B., D.A.L. and S.S.T.; writing—original draft preparation, V.A.A., P.A.Z., D.A.L., O.V.B., E.B.G., S.O.P., A.M.K. and Y.A.P.; writing—review and editing, V.A.A., S.S.T., P.A.Z., O.A.B., E.B.G. and L.I.E.; visualization, V.A.A., S.S.T., O.A.B. and E.B.G.; supervision, S.S.T. and P.V.S.; project administration, V.A.A. and S.S.T.; funding acquisition, S.S.T. All authors have read and agreed to the published version of the manuscript.

Funding: This work was supported by the Ministry of Science and Higher Education of the Russian Federation (Agreement No. 075-15-2021-1049).

Institutional Review Board Statement: Not applicable.

Informed Consent Statement: Not applicable.

Data Availability Statement: The data presented in this study are openly available in NCBI at accession number JBHFFO000000000.

Acknowledgments: We express our sincere gratitude to Ilya A. Osterman for his invaluable assistance in reviewing the manuscript and consulting on technical questions.

Conflicts of Interest: The authors declare no conflicts of interest.

References

1. Nathans, D. Puromycin Inhibition of Protein Synthesis: Incorporation of Puromycin into Peptide Chains. *Proc. Natl. Acad. Sci. USA* **1964**, *51*, 585–592. [CrossRef]
2. Pestka, S. Inhibitors of Ribosome Functions. *Annu. Rev. Microbiol.* **1971**, *25*, 487–562. [CrossRef]
3. Schmidt, E.K.; Clavarino, G.; Ceppi, M.; Pierre, P. SUnSET, a Nonradioactive Method to Monitor Protein Synthesis. *Nat. Methods* **2009**, *6*, 275–277. [CrossRef]
4. Tom Dieck, S.; Kochen, L.; Hanus, C.; Heumüller, M.; Bartnik, I.; Nassim-Assir, B.; Merk, K.; Mosler, T.; Garg, S.; Bunse, S.; et al. Direct Visualization of Newly Synthesized Target Proteins in Situ. *Nat. Methods* **2015**, *12*, 411–414. [CrossRef]
5. Aviner, R. The Science of Puromycin: From Studies of Ribosome Function to Applications in Biotechnology. *Comput. Struct. Biotechnol. J.* **2020**, *18*, 1074–1083. [CrossRef]

6. Scarpitti, M.; Kearse, M. In Vitro Analysis of Stalled Ribosomes Using Puromycin Incorporation. *Bio-Protocol* **2023**, *13*, E4744. [CrossRef]
7. David, A.; Dolan, B.P.; Hickman, H.D.; Knowlton, J.J.; Clavarino, G.; Pierre, P.; Bennink, J.R.; Yewdell, J.W. Nuclear Translation Visualized by Ribosome-Bound Nascent Chain Puromycylation. *J. Cell Biol.* **2012**, *197*, 45–57. [CrossRef]
8. Aviner, R.; Geiger, T.; Elroy-Stein, O. Genome-Wide Identification and Quantification of Protein Synthesis in Cultured Cells and Whole Tissues by Puromycin-Associated Nascent Chain Proteomics (PUNCH-P). *Nat. Protoc.* **2014**, *9*, 751–760. [CrossRef]
9. Seelig, B. mRNA Display for the Selection and Evolution of Enzymes from in Vitro-Translated Protein Libraries. *Nat. Protoc.* **2011**, *6*, 540–552. [CrossRef]
10. Rubio, M.A.; Espinosa, J.C.; Tercero, J.A.; Jiménez, A. The Pur10 Protein Encoded in the Gene Cluster for Puromycin Biosynthesis of *Streptomyces alboniger* Is an NAD-Dependent ATP Dehydrogenase. *FEBS Lett.* **1998**, *437*, 197–200. [CrossRef]
11. Tercero, J.A.; Espinosa, J.C.; Lacalle, R.A.; Jiménez, A. The Biosynthetic Pathway of the Aminonucleoside Antibiotic Puromycin, as Deduced from the Molecular Analysis of the Pur Cluster of *Streptomyces alboniger*. *J. Biol. Chem.* **1996**, *271*, 1579–1590. [CrossRef]
12. Ángel Rubio, M.; Barrado, P.; Carlos Espinosa, J.; Jiménez, A.; Lobato, M.F. The *Pur6* Gene of the Puromycin Biosynthetic Gene Cluster from *Streptomyces alboniger* Encodes a Tyrosinyl-Aminonucleoside Synthetase. *FEBS Lett.* **2004**, *577*, 371–375. [CrossRef]
13. Tercero, J.A.; Lacalle, R.A.; Jimenez, A. The *Pur8* Gene from the *Pur* Cluster of *Streptomyces alboniger* Encodes a Highly Hydrophobic Polypeptide Which Confers Resistance to Puromycin. *Eur. J. Biochem.* **1993**, *218*, 963–971. [CrossRef]
14. Lacalle, R.A.; Pulido, D.; Vara, J.; Zaiacaín, M.; Jiménez, A. Molecular Analysis of the *Pac* Gene Encoding a Puromycin N-Acetyl Transferase from *Streptomyces alboniger*. *Gene* **1989**, *79*, 375–380. [CrossRef]
15. Lacalle, R.A.; Tercero, J.A.; Vara, J.; Jimenez, A. Identification of the Gene Encoding an N-Acetylpuromycin N-Acetylhydrolase in the Puromycin Biosynthetic Gene Cluster from *Streptomyces alboniger*. *J. Bacteriol.* **1993**, *175*, 7474–7478. [CrossRef]
16. Tercero, J.A.; Espinosa, J.C.; Jiménez, A. Expression of the *Streptomyces alboniger pur* Cluster in *Streptomyces lividans* Is Dependent on the *bldA*-Encoded $tRNA^{Leu}$. *FEBS Lett.* **1998**, *421*, 221–223. [CrossRef]
17. Lacalle, R.A.; Ruiz, D.; Jiménez, A. Molecular Analysis of the *dmpM* Gene Encoding an O-Demethyl Puromycin O-Methyltransferase from *Streptomyces alboniger*. *Gene* **1991**, *109*, 55–61. [CrossRef]
18. Tercero, J.A.; Espinosa, J.C.; Jiménez, A. StgR, a New *Streptomyces alboniger* Member of the LysR Family of Transcriptional Regulators. *Mol. Gen. Genet.* **1998**, *259*, 475–483. [CrossRef]
19. Lacalle, R.A.; Tercero, J.A.; Jiménez, A. Cloning of the Complete Biosynthetic Gene Cluster for an Aminonucleoside Antibiotic, Puromycin, and Its Regulated Expression in Heterologous Hosts. *EMBO J.* **1992**, *11*, 785–792. [CrossRef]
20. Espinosa, J.C.; Tercero, J.A.; Rubio, M.A.; Jiménez, A. The *Pur7* Gene from the Puromycin Biosynthetic Pur Cluster of *Streptomyces alboniger* Encodes a Nudix Hydrolase. *J. Bacteriol.* **1999**, *181*, 4914–4918. [CrossRef]
21. Sánchez, M.B.; Barrado, P.; Jiménez, A.; Fernández Lobato, M. The *Pur3* Gene from the *Pur* Cluster Encodes a Monophosphatase Essential for Puromycin Biosynthesis in *Streptomyces*. *FEBS Lett.* **2006**, *580*, 1807–1811. [CrossRef]
22. Paik, S.-Y.; Sugiyama, M.; Nomi, R. Isolation and Properties of a Puromycin Acetyltransferase from Puromycin-Producing Streptomyces Alboniger. *J. Antibiot.* **1985**, *38*, 1761–1766. [CrossRef]
23. Vara, J.A.; Portela, A.; Ortìn, J.; Jimènez, A. Expression in Mammalian Cells of a Gene from *Streptomyces alboniger* Conferring Puromycin Resistance. *Nucl. Acids Res.* **1986**, *14*, 4617–4624. [CrossRef]
24. Kyung, S.-J.; Lee, H.-C.; Liou, K.-K.; Lee, E.-B.; Kang, S.-Y. Cystocin, a Novel Antibiotic, Produced by *Streptomyces* Sp. GCA0001: Production and Characterization of Cystocin. *J. Microbiol. Biotechnol.* **2003**, *13*, 483–486.
25. Chen, W.; Qi, J.; Wu, P.; Wan, D.; Liu, J.; Feng, X.; Deng, Z. Natural and Engineered Biosynthesis of Nucleoside Antibiotics in *Actinomycetes*. *J. Ind. Microbiol. Biotechnol.* **2016**, *43*, 401–417. [CrossRef]
26. Saugar, I.; Sanz, E.; Rubio, M.Á.; Espinosa, J.C.; Jiménez, A. Identification of a Set of Genes Involved in the Biosynthesis of the Aminonucleoside Moiety of Antibiotic A201A from *Streptomyces capreolus*: Aminonucleoside A201A Biosynthetic Genes. *Eur. J. Biochem.* **2002**, *269*, 5527–5535. [CrossRef]
27. Saugar, I.; Molloy, B.; Sanz, E.; Blanca Sánchez, M.; Fernández-Lobato, M.; Jiménez, A. Characterization of the Biosynthetic Gene Cluster (Ata) for the A201A Aminonucleoside Antibiotic from *Saccharothrix mutabilis* Subsp. *Capreolus*. *J. Antibiot.* **2017**, *70*, 404–413. [CrossRef]
28. Barrasa, M.I.; Tercero, J.A.; Jimenez, A. The Aminonucleoside Antibiotic A201A Is Inactivated by a Phosphotransferase Activity from *Streptomyces Capreolus* NRRL 3817, the Producing Organism. Isolation and Molecular Characterization of the Relevant Encoding Gene and Its DNA Flanking Regions. *Eur. J. Biochem.* **1997**, *245*, 54–63. [CrossRef]
29. Zhu, Q.; Li, J.; Ma, J.; Luo, M.; Wang, B.; Huang, H.; Tian, X.; Li, W.; Zhang, S.; Zhang, C.; et al. Discovery and Engineered Overproduction of Antimicrobial Nucleoside Antibiotic A201A from the Deep-Sea Marine Actinomycete *Marinactinospora Thermotolerans* Scsio 00652. *Antimicrob. Agents Chemother.* **2012**, *56*, 110–114. [CrossRef]
30. Shahsavari, N.; Wang, B.; Imai, Y.; Mori, M.; Son, S.; Liang, L.; Böhringer, N.; Manuse, S.; Gates, M.F.; Morrissette, M.; et al. A Silent Operon of *Photorhabdus luminescens* Encodes a Prodrug Mimic of GTP. *mBio* **2022**, *13*, e00700-22. [CrossRef]
31. August, P.R.; Tang, L.; Yoon, Y.J.; Ning, S.; Müller, R.; Yu, T.-W.; Taylor, M.; Hoffmann, D.; Kim, C.-G.; Zhang, X.; et al. Biosynthesis of the Ansamycin Antibiotic Rifamycin: Deductions from the Molecular Analysis of the Rif Biosynthetic Gene Cluster of *Amycolatopsis mediterranei* S699. *Chem. Biol.* **1998**, *5*, 69–79. [CrossRef]

32. Arakawa, K.; Müller, R.; Mahmud, T.; Yu, T.-W.; Floss, H.G. Characterization of the Early Stage Aminoshikimate Pathway in the Formation of 3-Amino-5-Hydroxybenzoic Acid: The RifN Protein Specifically Converts Kanosamine into Kanosamine 6-Phosphate. *J. Am. Chem. Soc.* **2002**, *124*, 10644–10645. [CrossRef]
33. Zhang, W.; Ntai, I.; Kelleher, N.L.; Walsh, C.T. tRNA-Dependent Peptide Bond Formation by the Transferase PacB in Biosynthesis of the Pacidamycin Group of Pentapeptidyl Nucleoside Antibiotics. *Proc. Natl. Acad. Sci. USA* **2011**, *108*, 12249–12253. [CrossRef]
34. Lauer, B.; Russwurm, R.; Schwarz, W.; Kálmánczhelyi, A.; Bruntner, C.; Rosemeier, A.; Bormann, C. Molecular Characterization of Co-Transcribed Genes from *Streptomyces tendae* Tü901 Involved in the Biosynthesis of the Peptidyl Moiety and Assembly of the Peptidyl Nucleoside Antibiotic Nikkomycin. *Mol. Gen. Genet.* **2001**, *264*, 662–673. [CrossRef]
35. Cone, M.C.; Yin, X.; Grochowski, L.L.; Parker, M.R.; Zabriskie, T.M. The Blasticidin s Biosynthesis Gene Cluster from *Streptomyces griseochromogenes*: Sequence Analysis, Organization, and Initial Characterization. *ChemBioChem* **2003**, *4*, 821–828. [CrossRef]
36. Li, J.; Li, L.; Tian, Y.; Niu, G.; Tan, H. Hybrid Antibiotics with the Nikkomycin Nucleoside and Polyoxin Peptidyl Moieties. *Metab. Eng.* **2011**, *13*, 336–344. [CrossRef]
37. Chen, W.; Huang, T.; He, X.; Meng, Q.; You, D.; Bai, L.; Li, J.; Wu, M.; Li, R.; Xie, Z.; et al. Characterization of the Polyoxin Biosynthetic Gene Cluster from *Streptomyces cacaoi* and Engineered Production of Polyoxin H. *J. Biol. Chem.* **2009**, *284*, 10627–10638. [CrossRef]
38. Fawaz, M.V.; Topper, M.E.; Firestine, S.M. The ATP-Grasp Enzymes. *Bioorganic Chem.* **2011**, *39*, 185–191. [CrossRef]
39. Grammel, N.; Pankevych, K.; Demydchuk, J.; Lambrecht, K.; Saluz, H.-P.; Keller, U.; Krügel, H. A β-Lysine Adenylating Enzyme and a β-Lysine Binding Protein Involved in Poly β-Lysine Chain Assembly in Nourseothricin Synthesis in *Streptomyces noursei*: Poly β-Lysine Assembly in *S. noursei*. *Eur. J. Biochem.* **2002**, *269*, 347–357. [CrossRef]
40. Niu, G.; Li, L.; Wei, J.; Tan, H. Cloning, Heterologous Expression, and Characterization of the Gene Cluster Required for Gougerotin Biosynthesis. *Chem. Biol.* **2013**, *20*, 34–44. [CrossRef]
41. Osterman, I.A.; Komarova, E.S.; Shiryaev, D.I.; Korniltsev, I.A.; Khven, I.M.; Lukyanov, D.A.; Tashlitsky, V.N.; Serebryakova, M.V.; Efremenkova, O.V.; Ivanenkov, Y.A.; et al. Sorting Out Antibiotics' Mechanisms of Action: A Double Fluorescent Protein Reporter for High-Throughput Screening of Ribosome and DNA Biosynthesis Inhibitors. *Antimicrob. Agents Chemother.* **2016**, *60*, 7481–7489. [CrossRef]
42. Abbas, M.; Elshahawi, S.I.; Wang, X.; Ponomareva, L.V.; Sajid, I.; Shaaban, K.A.; Thorson, J.S. Puromycins B–E, Naturally Occurring Amino-Nucleosides Produced by the Himalayan Isolate *Streptomyces* Sp. PU-14G. *J. Nat. Prod.* **2018**, *81*, 2560–2566. [CrossRef]
43. Caputo, A.T.; Eder, O.M.; Bereznakova, H.; Pothuis, H.; Ardevol, A.; Newman, J.; Nuttall, S.; Peat, T.S.; Adams, T.E. Structure-Guided Selection of Puromycin N-Acetyltransferase Mutants with Enhanced Selection Stringency for Deriving Mammalian Cell Lines Expressing Recombinant Proteins. *Sci. Rep.* **2021**, *11*, 5247. [CrossRef]
44. Altschul, S.F.; Gish, W.; Miller, W.; Myers, E.W.; Lipman, D.J. Basic Local Alignment Search Tool. *J. Mol. Biol.* **1990**, *215*, 403–410. [CrossRef]
45. Blin, K.; Shaw, S.; Augustijn, H.E.; Reitz, Z.L.; Biermann, F.; Alanjary, M.; Fetter, A.; Terlouw, B.R.; Metcalf, W.W.; Helfrich, E.J.N.; et al. antiSMASH 7.0: New and Improved Predictions for Detection, Regulation, Chemical Structures and Visualisation. *Nucleic Acids Res.* **2023**, *51*, W46–W50. [CrossRef]
46. Katoh, K.; Rozewicki, J.; Yamada, K.D. MAFFT Online Service: Multiple Sequence Alignment, Interactive Sequence Choice and Visualization. *Brief. Bioinform.* **2019**, *20*, 1160–1166. [CrossRef]
47. Gilchrist, C.L.M.; Chooi, Y.-H. Clinker & Clustermap.Js: Automatic Generation of Gene Cluster Comparison Figures. *Bioinformatics* **2021**, *37*, 2473–2475. [CrossRef]
48. Lowe, T.M.; Chan, P.P. tRNAscan-SE On-Line: Integrating Search and Context for Analysis of Transfer RNA Genes. *Nucleic Acids Res.* **2016**, *44*, W54–W57. [CrossRef]
49. Brown, N.P.; Leroy, C.; Sander, C. MView: A Web-Compatible Database Search or Multiple Alignment Viewer. *Bioinformatics* **1998**, *14*, 380–381. [CrossRef]
50. Abramson, J.; Adler, J.; Dunger, J.; Evans, R.; Green, T.; Pritzel, A.; Ronneberger, O.; Willmore, L.; Ballard, A.J.; Bambrick, J.; et al. Accurate Structure Prediction of Biomolecular Interactions with AlphaFold 3. *Nature* **2024**, *630*, 493–500. [CrossRef]
51. Kolmogorov, M.; Yuan, J.; Lin, Y.; Pevzner, P.A. Assembly of Long, Error-Prone Reads Using Repeat Graphs. *Nat. Biotechnol.* **2019**, *37*, 540–546. [CrossRef] [PubMed]
52. Walker, B.J.; Abeel, T.; Shea, T.; Priest, M.; Abouelliel, A.; Sakthikumar, S.; Cuomo, C.A.; Zeng, Q.; Wortman, J.; Young, S.K.; et al. Pilon: An Integrated Tool for Comprehensive Microbial Variant Detection and Genome Assembly Improvement. *PLoS ONE* **2014**, *9*, e112963. [CrossRef] [PubMed]
53. Seemann, T. Prokka: Rapid Prokaryotic Genome Annotation. *Bioinformatics* **2014**, *30*, 2068–2069. [CrossRef]
54. Medema, M.H.; Takano, E.; Breitling, R. Detecting Sequence Homology at the Gene Cluster Level with MultiGeneBlast. *Mol. Biol. Evol.* **2013**, *30*, 1218–1223. [CrossRef]
55. Kautsar, S.A.; Blin, K.; Shaw, S.; Navarro-Muñoz, J.C.; Terlouw, B.R.; van der Hooft, J.J.J.; van Santen, J.A.; Tracanna, V.; Suarez Duran, H.G.; Pascal Andreu, V.; et al. MIBiG 2.0: A Repository for Biosynthetic Gene Clusters of Known Function. *Nucleic Acids Res.* **2020**, *48*, D454–D458. [CrossRef]
56. Yoon, S.-H.; Ha, S.-M.; Kwon, S.; Lim, J.; Kim, Y.; Seo, H.; Chun, J. Introducing EzBioCloud: A Taxonomically United Database of 16S rRNA Gene Sequences and Whole-Genome Assemblies. *Int. J. Syst. Evol. Microbiol.* **2017**, *67*, 1613–1617. [CrossRef] [PubMed]

57. Goris, J.; Konstantinidis, K.T.; Klappenbach, J.A.; Coenye, T.; Vandamme, P.; Tiedje, J.M. DNA–DNA Hybridization Values and Their Relationship to Whole-Genome Sequence Similarities. *Int. J. Syst. Evol. Microbiol.* **2007**, *57*, 81–91. [CrossRef]
58. Konstantinidis, K.T.; Ramette, A.; Tiedje, J.M. Toward a More Robust Assessment of Intraspecies Diversity, Using Fewer Genetic Markers. *Appl. Environ. Microbiol.* **2006**, *72*, 7286–7293. [CrossRef] [PubMed]
59. Hu, S.; Li, K.; Zhang, Y.; Wang, Y.; Fu, L.; Xiao, Y.; Tang, X.; Gao, J. New Insights Into the Threshold Values of Multi-Locus Sequence Analysis, Average Nucleotide Identity and Digital DNA–DNA Hybridization in Delineating Streptomyces Species. *Front. Microbiol.* **2022**, *13*, 910277. [CrossRef]
60. Chun, J.; Oren, A.; Ventosa, A.; Christensen, H.; Arahal, D.R.; Da Costa, M.S.; Rooney, A.P.; Yi, H.; Xu, X.-W.; De Meyer, S.; et al. Proposed Minimal Standards for the Use of Genome Data for the Taxonomy of Prokaryotes. *Int. J. Syst. Evol. Microbiol.* **2018**, *68*, 461–466. [CrossRef]
61. Meier-Kolthoff, J.P.; Göker, M. TYGS Is an Automated High-Throughput Platform for State-of-the-Art Genome-Based Taxonomy. *Nat. Commun.* **2019**, *10*, 2182. [CrossRef] [PubMed]
62. Richter, M.; Rosselló-Móra, R.; Oliver Glöckner, F.; Peplies, J. JSpeciesWS: A Web Server for Prokaryotic Species Circumscription Based on Pairwise Genome Comparison. *Bioinformatics* **2016**, *32*, 929–931. [CrossRef] [PubMed]
63. Krassilnikov, N.A.; Koveshnikov, A.D. *Actinomyces tumemacerans* n. Sp.—A New Species Inducing Disintegration of Tumors in Plant. *Mikrobiologiia* **1962**, *31*, 589–594.
64. Pridham, T.G. New Names and New Combinations in the Order Actinomycetales Buchanan 1917. *Bull. U.S. Dep. Agric.* **1970**, *1424*, 1–55.
65. Terekhov, S.S.; Smirnov, I.V.; Malakhova, M.V.; Samoilov, A.E.; Manolov, A.I.; Nazarov, A.S.; Danilov, D.V.; Dubiley, S.A.; Osterman, I.A.; Rubtsova, M.P.; et al. Ultrahigh-Throughput Functional Profiling of Microbiota Communities. *Proc. Natl. Acad. Sci. USA* **2018**, *115*, 9551–9556. [CrossRef]
66. Baba, T.; Ara, T.; Hasegawa, M.; Takai, Y.; Okumura, Y.; Baba, M.; Datsenko, K.A.; Tomita, M.; Wanner, B.L.; Mori, H. Construction of *Escherichia coli* K-12 In-frame, Single-gene Knockout Mutants: The Keio Collection. *Mol. Syst. Biol.* **2006**, *2*, 2006.0008. [CrossRef] [PubMed]
67. Riss, T.L.; Moravec, R.A.; Niles, A.L.; Duellman, S.; Benink, H.A.; Worzella, T.J.; Minor, L. Cell Viability Assays. In *Assay Guidance Manual*; Markossian, S., Grossman, A., Arkin, M., Auld, D., Austin, C., Baell, J., Brimacombe, K., Chung, T.D.Y., Coussens, N.P., Dahlin, J.L., et al., Eds.; Eli Lilly & Company and the National Center for Advancing Translational Sciences: Bethesda, MD, USA, 2004.
68. Marina, V.I.; Bidzhieva, M.; Tereshchenkov, A.G.; Orekhov, D.; Sagitova, V.E.; Sumbatyan, N.V.; Tashlitsky, V.N.; Ferberg, A.S.; Maviza, T.P.; Kasatsky, P.; et al. An Easy Tool to Monitor the Elemental Steps of In Vitro Translation via Gel Electrophoresis of Fluorescently Labeled Small Peptides. *RNA* **2024**, *30*, 298–307. [CrossRef]
69. Orelle, C.; Carlson, S.; Kaushal, B.; Almutairi, M.M.; Liu, H.; Ochabowicz, A.; Quan, S.; Pham, V.C.; Squires, C.L.; Murphy, B.T.; et al. Tools for Characterizing Bacterial Protein Synthesis Inhibitors. *Antimicrob. Agents Chemother.* **2013**, *57*, 5994–6004. [CrossRef]

International Journal of
Molecular Sciences

MDPI

Article

Gut Microbiota and Inflammation Modulation in a Rat Model for Ulcerative Colitis after the Intraperitoneal Administration of Apigenin, Luteolin, and Xanthohumol

Patricia Magadán-Corpas [1,2,3], Álvaro Pérez-Valero [1,2,3], Suhui Ye [1,2,3], Sandra Sordon [4], Ewa Huszcza [4], Jarosław Popłoński [4], Claudio J. Villar [1,2,3] and Felipe Lombó [1,2,3,*]

1 Research Group BIONUC (Biotechnology of Nutraceuticals and Bioactive Compounds), Departamento de Biología Funcional, Área de Microbiología, Universidad de Oviedo, 33006 Oviedo, Spain; magadanpatricia@uniovi.es (P.M.-C.); apv.moratalla@gmail.com (Á.P.-V.); yesuhui@uniovi.es (S.Y.); cjvg@uniovi.es (C.J.V.)

2 IUOPA (Instituto Universitario de Oncología del Principado de Asturias), 33006 Oviedo, Spain

3 ISPA (Instituto de Investigación Sanitaria del Principado de Asturias), 33006 Oviedo, Spain

4 Department of Food Chemistry and Biocatalysis, Wrocław University of Environmental and Life Sciences, Norwida 25, 50-375 Wrocław, Poland; sandra.sordon@upwr.edu.pl (S.S.); ewa.huszcza@upwr.edu.pl (E.H.); jaroslaw.poplonski@upwr.edu.pl (J.P.)

* Correspondence: lombofelipe@uniovi.es; Tel.: +34-985103593

Abstract: Ulcerative colitis (UC) is a chronic inflammatory disorder affecting the colon, with symptomatology influenced by factors including environmental, genomic, microbial, and immunological interactions. Gut microbiota dysbiosis, characterized by bacterial population alterations, contributes to intestinal homeostasis disruption and aberrant immune system activation, thereby exacerbating the inflammatory state. This study assesses the therapeutic efficacy of intraperitoneal (IP) injected flavonoids (apigenin, luteolin, and xanthohumol) in the reduction of inflammatory parameters and the modulation of the gut microbiota in a murine model of ulcerative colitis. Flavonoids interact with gut microbiota by modulating their composition and serving as substrates for the fermentation into other anti-inflammatory bioactive compounds. Our results demonstrate the effectiveness of luteolin and xanthohumol treatment in enhancing the relative abundance of anti-inflammatory microorganisms, thereby attenuating pro-inflammatory species. Moreover, all three flavonoids exhibit efficacy in the reduction of pro-inflammatory cytokine levels, with luteolin strongly demonstrating utility in alleviating associated physical UC symptoms. This suggests that this molecule is a potential alternative or co-therapy to conventional pharmacological interventions, potentially mitigating their adverse effects. A limited impact on microbiota is observed with apigenin, and this is attributed to its solubility constraints via the chosen administration route, resulting in its accumulation in the mesentery.

Keywords: inflammatory bowel disease; gut microbiota; flavonoid; anti-inflammatory

1. Introduction

Inflammatory bowel disease (IBD) is a term that encompasses two conditions, Crohn's disease (CD) and ulcerative colitis (UC), both characterized by chronic inflammation of the gastrointestinal tract. While CD can affect any part of the intestinal tract (from mouth to anus), UC only affects the colon, causing abdominal pain, mucus, diarrhea, and blood stool [1]. IBD represents a group of archetypal complex disorders distinguished by chronic and varied symptoms, influenced by the interplay among environmental, genomic, microbial, and immunological factors [2].

Multiple investigations have corroborated notable distinctions in the composition, diversity, and/or abundance of gut microbiota (dysbiosis) between healthy individuals and those with IBD, leading to the loss of intestinal homeostasis or improper immune activation [3–5]. A lot of research is being conducted to better understand how the composition of the gut microbiota may influence the development and progression of IBD and existing common microbial signatures shared among patients with this pathology, such as an increase in the phylum *Pseudomonadota* [6,7]. The gut microbiome dysbiosis present in IBD patients is closely related to inflammation as it can increase the expression of inflammatory cytokines by the intestinal T lymphocytes. Several mediators contribute to the development of this chronic intestinal inflammation, with the primary ones being the interleukins IL-1β, IL-6, and IL-17 [8–11]. Additionally, the disruption of the mucosal barrier associated with IBD leads to alterations in the taxa composition of mucus communities. Commensal microorganisms may transition into pathogenic entities (pathobionts), thereby initiating and perpetuating the inflammatory process through aberrant activation of the mucosal immune system [12–14]. Furthermore, the production of short-chain fatty acids (SCFAs), generated from the metabolism of certain microorganisms on the mucus layer and dietary prebiotic carbohydrates, is also compromised. These SCFAs are associated with anti-inflammatory properties, the maintenance of normal mucosal function, and the regulation of intestinal immune homeostasis [12,15–17]. It has not yet been possible to determine whether the dysbiosis associated with IBD is the cause or a consequence of this pathology. Nevertheless, it is certain that the deregulation of the gut microbiota balance contributes to the evolution and progression of the IBD, supporting and maintaining the inflammatory responses.

At the clinical level, UC requires long-term potent pharmacological treatment [18], and it may be difficult to find a suitable medication without serious side effects. Conversely, flavonoids constitute a family of natural polyphenolic compounds derived from plants, which have been previously demonstrated to have anti-inflammatory properties. In this work, we have evaluated the effectiveness of three intraperitoneal (IP) injected flavonoids (apigenin, luteolin, and xanthohumol) in the amelioration of the symptoms associated with a UC rat model after dextran sodium sulfate (DSS) induction, as well as their influence on the gut microbiota composition. These three flavonoids have already been proven as antitumor compounds in previous in vitro experiments with human colon cancer cell lines [19], and their effectiveness in the treatment of IBD after oral administration has been shown in vivo [1,20–23]. Meanwhile, their therapeutic potential through IP administration remains an area warranting further comprehensive investigation to encompass the direct effects of flavonoids. This approach circumvents the initial gastrointestinal (GI) transit, thereby avoiding processes of degradation and modification of these biopharmaceutical compounds [24]. Studies also reported IP administration of small molecule pharmacological agents with a faster and more complete absorption, compared to oral routes [25]. Furthermore, a diminished dosage of flavonoids is required.

In addition, the impact of xanthohumol on the gut microbiota remains unexplored. In this work, we have characterized, for the first time, the gut microbiota changes induced by xanthohumol in a UC murine model.

The primary aim of this research was to assess the potential therapeutic efficacy of IP administration of flavonoids as a novel and potential therapeutic intervention for UC in a rat model of this disease. In pursuit of this goal, multiple parameters relevant to the progression of UC have been evaluated, including gut microbiota characterization. This approach has encompassed both the direct effects of flavonoids and their indirect effects mediated through the modulation of gut microbiota populations. The changes in the gut microbiota composition caused by the flavonoid treatments have been associated with specific bacterial taxa. These alterations in the bacterial populations may therefore modify the metabolic pathways of the production of bioactive metabolites [26,27]. Finally, these breakdown metabolites (e.g., SCFAs, aromatic flavonoids breakdown derivatives, etc.) may act at different levels in the colon mucosa, downregulating the inflammatory status (e.g., at the level of lymphocyte population modulation, myeloperoxidase, cytokines, barrier function, etc.)

The experiments described in this work demonstrate that luteolin, and to a lesser extent xanthohumol, are successful in the treatment of UC via IP injection. Our main findings are that both flavonoids exert their activity over the modulation of the gut microbiota community structure towards a decrease in pro-inflammatory taxa and an increase in anti-inflammatory taxa. Some of these anti-inflammatory taxa are well-known producers of beneficial compounds, such as SCFAs and flavonoid metabolites, which may further contribute to modulating pro-inflammatory cytokines in the intestinal mucosa. Statistically significant differences have been observed in the reduction of the cytokines IL-6 and IL-1β in the three flavonoid treatment cohorts, likely linked to either the direct action of flavonoids or the metabolites resulting from the gut microbiota modulation. Finally, the luteolin observed effects are further extended to an improvement in colon ulceration and stool consistency parameters. However, in this study, IP administration has caused a reduced bioavailability of apigenin (and reduced protection against UC) due to the presence of precipitate granules in the mesentery, a fact that had not been previously described, and is probably derived from its lower hydrophilicity (in comparison with luteolin).

2. Results

Comparisons of the analyzed parameters were conducted between the PBS (phosphate buffer saline) control cohort, serving as the disease model, and each of the cohorts receiving flavonoid treatments. The analyzed parameters encompassed daily body weight measurements, serving as an indicator of effective digestive function as animals with UC typically exhibit compromised nutrient absorption and consequent reductions in body weight. Additionally, the disease activity index (DAI), which integrates changes in body weight and stool consistency (including diarrhea severity and the presence of blood in feces) as clinical indicators of disease severity, was assessed. Following euthanasia at the conclusion of the experiment, three macroscopic histological parameters associated with inflammation were investigated: hyperplastic Peyer's patches, which are lymphoid tissues in the small intestine that undergo macroscopic enlargement in response to inflammation; colon length, which typically diminishes during inflammatory conditions; and colon ulceration, indicative of mucosal alteration. Also linked to the inflammatory status, two types of cytokines (IL-1β and IL-6) were measured in plasma samples, and the myeloperoxidase concentration was evaluated in the colon mucosa (after tissue homogenization). Subsequently, the weight of the caecum was quantified, and bacterial populations within this organ were analyzed using 16S rRNA next-generation sequencing (NGS) on an Illumina platform.

2.1. Effect of Flavonoid Treatments on Body Weight and DAI

Body weight was monitored daily throughout the entire experimental period. Animals induced with UC exhibited discernible patterns in the comparison between cohorts (Figure 1). Notably, there was an initial increase in body weight until day 9, followed by a subsequent decline in all the cohorts, attributed to the induction of UC. It is noteworthy that this decline in body weight was particularly pronounced in the UC-induced animals from the non-treated (phosphate-buffered saline, PBS) and apigenin-treated cohorts, reaching its limit at day 11. All cohorts exhibited a recuperation phase toward the conclusion of the experiment as the effects of DSS treatment gradually waned.

By the end of the experiment (day 12), the mean body weight value for the UC-induced animals in the PBS cohort was 139.7 ± 15.9 g, while the two healthy absolute control animals reached 181.3 ± 20.1 g on average. Regarding the apigenin, luteolin, and xanthohumol cohorts, the mean weights at day 12 in the case of the UC-induced animals were 143.8 ± 19.4 g, 150.2 ± 8.3 g, and 149.6 ± 18.6 g, respectively, while their corresponding healthy controls weights were 183.4 ± 9.3 g, 170.6 ± 3 g, and 154.4 ± 17.1. Statistical analyses showed no significant differences between the PBS and either of the flavonoid treatment cohorts regarding body weight (Supplementary Figure S1a).

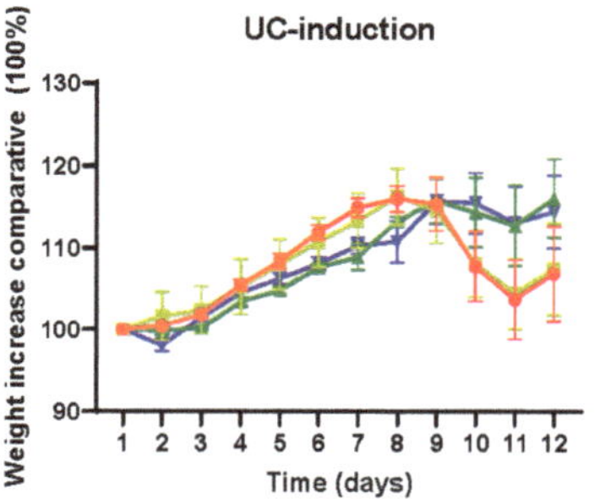

Figure 1. Comparison of the body weight increase in UC-induced animals from the four studied cohorts. The reductions in body weight observed around days 9 to 11 in UC-induced animals are due to the peak in UC symptoms (reduced feed ingest due to colon inflammation). Note that this reduction was minimal in luteolin and xanthohumol cohorts, thus showing a protective effect exerted by these treatments. PBS (red); apigenin (yellow); luteolin (green); xanthohumol (blue).

The mean DAI scores on day 12 are represented in Figure 2a, exclusively for the UC-induced animal groups. A statistically significant difference was observed when comparing the luteolin-treated cohort to the control PBS cohort. Given the absence of statistically significant variations in body weight across cohorts, the enhancement in the DAI index in the luteolin cohort primarily stemmed from an amelioration in stool consistency (Figure 2b), one of the contributing parameters for DAI computation.

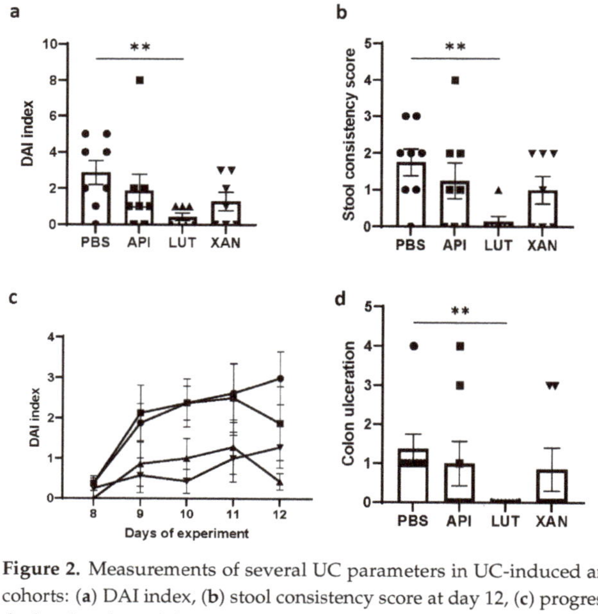

Figure 2. Measurements of several UC parameters in UC-induced animals from the four studied cohorts: (**a**) DAI index, (**b**) stool consistency score at day 12, (**c**) progression of the DAI index during the last five days of the experiment, and (**d**) colon ulceration at day 12. Comparisons were performed between the PBS control cohort and each flavonoid treatment cohort. In all cases, luteolin treatment showed a strong protective effect. Circle: control cohort; square: apigenin treatment cohort; upward triangle: luteolin treatment cohort; downward triangle: xanthohumol treatment cohort. Asterisks indicate statistically significant differences (** $p < 0.005$).

Furthermore, DAI index trends over the final five days of this study (Figure 2c) revealed a reduction in DAI scores within the luteolin and xanthohumol treatment cohorts, consistently maintaining lower scores compared to those observed in the PBS and apigenin cohorts during the peak period of UC induction.

2.2. Effect of Flavonoid Treatments on Hyperplastic Peyer's Patches, Colon Length, and Colon Ulceration

Peyer's patches were macroscopically quantified along the small intestine after euthanasia as this lymphoid tissue becomes hyperplastic in response to inflammatory processes, showing rounded, protruding, white 2–3 mm ovals on the surface of the small intestine [28]. Their reduction in number could then be associated with lower inflammatory signals. A statistically significant lower number of hyperplastic Peyer's patches was not observed in the flavonoid-treated animals compared to the PBS cohort regarding the UC-induced animals (Supplementary Figure S1b).

Colon length was also measured in all the surviving rats, as an indicator of colitis severity: the shorter the colon, the more severe the colitis inflammation. Again, no statistically significant differences were observed in the UC-induced animals between the PBS cohort and each of the three treatment cohorts (Supplementary Figure S1c).

Finally, colon ulceration was quantified for each individual within this study, and the findings are represented in Figure 2d. A statistically significant disparity was observed in the reduction of colon ulceration between the group of animals that were induced with UC and received luteolin treatment and the untreated UC-induced animals from the PBS cohort. Notably, all animals within the luteolin cohort displayed a complete absence of ulceration.

2.3. Effect of Flavonoid Treatments on Pro-Inflammatory Cytokines and Myeloperoxidase (MPO)

Statistical analyses revealed strong significant reductions in the pro-inflammatory cytokine IL-6 levels across all three flavonoid treatment groups compared to the PBS control cohort in the animals under UC induction. However, when examining cytokine IL-1β levels, xanthohumol did not exhibit statistically significant differences compared to the PBS control group, while apigenin and luteolin exhibited significant differences (Figure 3a,b).

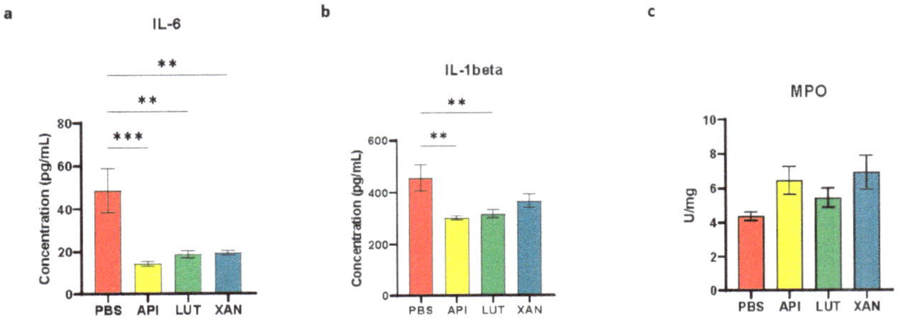

Figure 3. Measure of different pro-inflammatory biomarkers in UC-induced animals from the four studied cohorts: (**a**) IL-6, (**b**) IL-1β, and (**c**) MPO. Comparisons were performed between the PBS control cohort and each flavonoid treatment cohort. All three flavonoids demonstrated efficacy in reducing the levels of the pro-inflammatory cytokine IL-6, whereas only apigenin and luteolin exhibited effectiveness in modulating IL-1β. In contrast, flavonoids did not affect levels of MPO. PBS (red); apigenin (API, yellow); luteolin (LUT, green); xanthohumol (XAN, blue). Asterisks indicate statistically significant differences (** $p < 0.005$, *** $p < 0.0005$).

With respect to the enzymatic biomarker associated with colon mucosa inflammation (MPO), no statistically significant variances were detected in MPO tissue levels for either of the flavonoid treatments compared to the PBS control cohort (Figure 3c).

2.4. Caecum Weight and Metataxonomic Analyses of the Gut Microbiota

The caecum weight was assessed in the 38 surviving rats across the four cohorts, encompassing rats induced with UC. Statistical analysis revealed no significant differences in caecum weight in the comparison between the PBS cohort and each of the three treatment cohorts. (Supplementary Figure S1d).

In consideration of the microbiota studies, to assess the impact of flavonoid treatments on gut microbiota modulation subsequent to UC induction with DSS, comparisons were conducted on the composition of gut microbiota between the control PBS cohort and each specific flavonoid treatment cohort. Gut microbiota composition determination involved conducting a metataxonomics analysis of cecal content, utilizing 16S ribosomal RNA sequencing.

The two alpha diversity metrics, richness (observed OTUs: Operational Taxonomic Units) and evenness (a parameter that measures how numerically equal the bacterial community is in an experimental cohort, regarding the abundance and numbers of taxa), were measured within microbial communities, and alpha diversity was evaluated through the indices Chao1, Simpson, and Shannon. Boxplot representations of these indices are shown in Supplementary Figure S2. No statistically significant differences were found for any of these metrics between the UC-induced animals from the control PBS cohort and the flavonoid treatment cohorts, indicating no changes in terms of microbial alpha diversity.

The unweighted Unifrac beta diversity index (a qualitative parameter that measures the structural composition of bacterial communities between experimental animal cohorts, including the taxonomy data) was also calculated to evaluate differences between groups in terms of species complexity. The principal coordinate analysis (PCoA) plot for the visualization of microbial communities' structure is shown in Figure 4a. The analysis of beta diversity revealed notable statistically significant distinctions. Specifically, significant dissimilarities were discerned when contrasting the control PBS cohort with both the luteolin and xanthohumol cohorts (Figure 4b). Conversely, no statistically significant variations were evident when comparing the PBS cohort with the apigenin cohort. Furthermore, statistically significant disparities were also identified when comparing the apigenin cohort with both the luteolin and xanthohumol cohorts. However, there were no statistically significant distinctions between the luteolin and xanthohumol cohorts (Figure 4b).

a

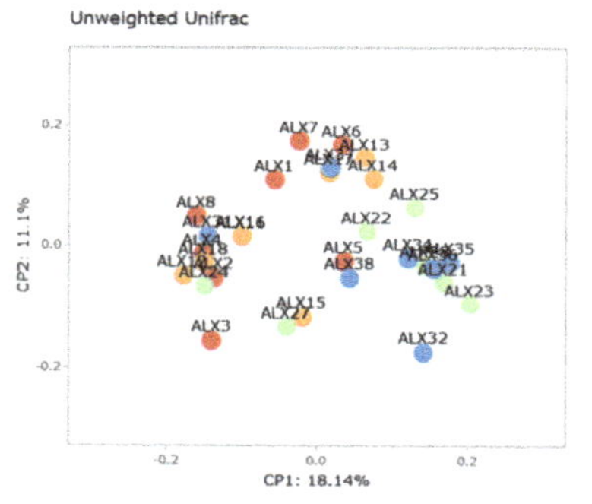

b

Unweigthed Unifrac	PBS	Apigenin	Luteolin
Apigenin			
Luteolin	*	*	
Xanthohumol	*	*	

Figure 4. Comparisons of gut microbiota beta diversity metrics between the four studied cohorts: (**a**) unweighted Unifrac PCoA plot and (**b**) Permanova test for the unweighted Unifrac beta diversity measure. As ascertained by the unweighted Unifrac beta diversity analysis, the animals within the luteolin and xanthohumol cohorts exhibited greater similarity to each other, while they were more distinct from both the control (PBS) and apigenin cohorts. Conversely, the control and apigenin cohorts demonstrated similarity in terms of microbial community structure. In the PCoA plot, each dot represents one animal, and distances between dots represent the ecological distances between samples. PBS (red); apigenin (yellow); luteolin (green); xanthohumol (blue). Asterisks indicate statistically significant differences (* $p < 0.05$).

The metataxonomics analysis of the UC-induced animals' gut microbiota showed statistically significant differences at different taxonomic levels and between different cohorts. In general, *Bacillota* and *Bacteroidota* constituted the most predominant phyla (90%) in all cohorts. The relative abundance of the other phyla varied depending on the treatment cohort (Figure 5, Table 1).

When the UC-induced animals from the PBS cohort were compared to the UC-induced animals from each flavonoid treatment cohort (Figure 5, Table 1), statistically significant differences were observed, with a significant reduction in the phylum *Bacillota* in the apigenin cohort (45.49% vs. 51.20% in the PBS animals), while the phylum *Bacteroidota* was significantly decreased in the xanthohumol cohort (34.99% vs. 40.05% in the PBS cohort). A reduction in the phylum *Pseudomonadota* was observed in the luteolin (0.23%) and xanthohumol (0.17%) cohorts with respect to the control cohort (1.28%). The phylum *Desulfobacterota*, with the genus *Bilophila* as its unique member, was also significantly decreased in the xanthohumol cohort (0.03% vs. 0.36% in the PBS animals). Also remarkable, and statistically significant, was the increase in the phylum *Verrucomicrobiota*, with *Akkermansia muciniphila* as its unique representative species in the xanthohumol cohort (9.77% vs. 4.56% in the PBS cohort).

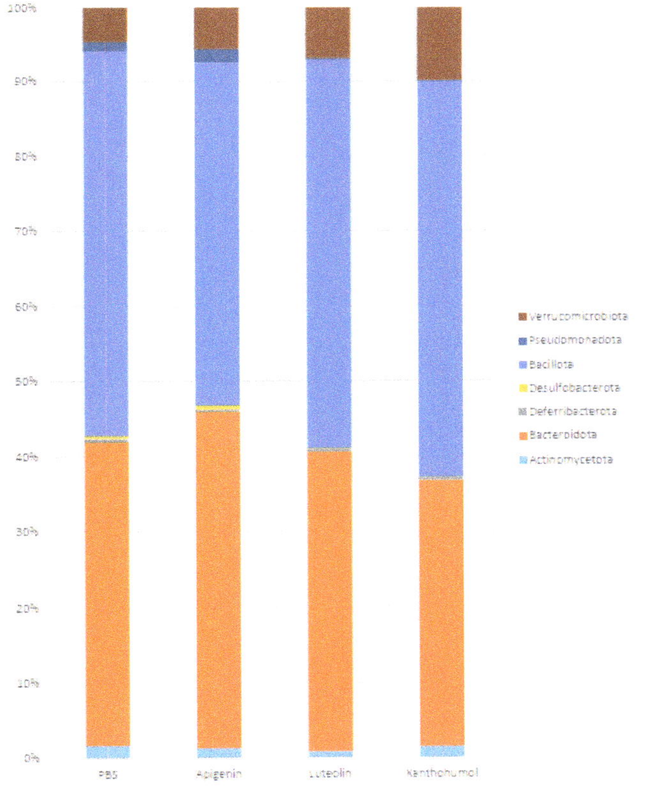

Figure 5. Barplot representation of the relative abundance (%) at the phylum level in the gut microbiota between animals induced with UC across the four experimental cohorts. *Bacillota* and *Bacteroidota* constitute 90% of the relative abundance in all cohorts. Higher differences between PBS and flavonoid treatment cohorts could be observed regarding the phyla *Pseudomonadota* and *Verrucomicrobiota*.

Table 1. Average relative abundance (%) at the phylum level in the gut microbiota between animals induced with UC, across the four experimental cohorts. The statistically significant differences in the comparisons between the control PBS cohort and each flavonoid treatment cohort are depicted. API: apigenin; LUT: luteolin; XAN: xanthohumol. Asterisks indicate statistically significant differences (* $p < 0.05$; ** $p < 0.005$; *** $p < 0.0005$).

Phylum	PBS	Apigenin	Luteolin	Xanthohumol	PBS vs. API	PBS vs. LUT	PBS vs. XAN
Actinomycetota	1.65	1.31	0.83	1.45			
Bacteroidota	40.05	44.32	39.40	34.99			*
Deferribacterota	0.46	0.29	0.33	0.35			
Desulfobacterota	0.36	0.53	0.07	0.03			*
Bacillota	51.20	45.49	51.48	52.32	*		
Pseudomonadota	1.28	1.87	0.23	0.17		**	***
Verrucomicrobiota	4.56	5.57	6.80	9.77			*

Regarding the comparisons between the three flavonoid treatment cohorts, statistically significant differences were observed in the case of the phyla *Bacteroidota*, *Desulfobacterota*, *Bacillota*, and *Pseudomonadota* between the apigenin and the xanthohumol cohorts. Luteolin showed a statistically significant reduction in comparison with apigenin regarding phylum *Pseudomonadota*.

The main differences at the family level in the comparison between the UC-induced animals from the PBS cohort and those treated with flavonoids were observed after luteolin and xanthohumol administration (Figure 6, Table 2, Supplementary Figure S3). The phylum *Bacillota* was significantly reduced only in the apigenin cohort. However, major changes were observed at the family level regarding the luteolin and xanthohumol cohorts. A statistically significant reduction was observed in the animals from all three flavonoid treatment cohorts regarding the families *Erysipelotrichaceae* (0.53% in the apigenin cohort, 0.38% in the luteolin cohort, and 0.36% in the xanthohumol cohort vs. 2.08% in the PBS cohort), *Streptococcaceae* (0.02%, 0.003%, and 0.02% vs. 0.16% in the PBS cohort) and *Staphylococcaceae* (genus *Staphylococcus*) (0.05%, 0, and 0.003% vs. 0.35% in the PBS cohort). The family *Peptococcaceae* showed a statistically significant reduction in the xanthohumol cohort (0.17% vs. 0.39% in the PBS cohort). Also statistically significant were the reductions in the luteolin and xanthohumol cohorts regarding the families *Anaerovoracaceae* (0.13% and 0.11%, respectively, vs. 0.33% in the PBS cohort), *Peptostreptococcaceae* (genus *Romboutsia*) (0.21% and 0.86% vs. 2.88% in the PBS cohort), and *Clostridiaceae* (genus *Clostridium sensu stricto 1*) (0.11% and 0.12% vs. 0.40% in the PBS cohort). Conversely, the family *Lachnospiraceae* was significantly increased in the luteolin cohort (23.25% vs. 14.96% in the PBS cohort). The decrease in the phylum *Bacteroidota* observed in the xanthohumol cohort could be mainly attributed to a reduction in the family *Rikenellaceae* (0.46% vs. 7.76% in the PBS cohort). A statistically significant increase was observed in the family *Prevotellaceae* in the xanthohumol cohort (3.83% vs. 1.73% in the PBS cohort). Two families from phylum *Pseudomonadota* were reduced in the UC animals from the luteolin and xanthohumol cohorts: *Sutterellaceae* (0.11% and 0.08%, respectively, vs. 0.54% in the PBS cohort) and *Enterobacteriaceae* (0.05% and 0.03%, respectively, vs. 0.68% in the PBS cohort) (Supplementary Figure S3). In the luteolin cohort, the family *Bifidobacteriaceae* (genus *Bifidobacterium*) (0.03% vs. 0.52% in the PBS cohort) showed a reduction (Supplementary Figure S3).

At the genus level (Table 3, Supplementary Figure S4), most of the differences were observed in the luteolin and xanthohumol cohorts compared to the PBS cohort. The genera *Turicibacter* and *Streptococcus*, as the most abundant ones of their respective families (*Erysipelotrichaceae* and *Streptococcaceae*), showed a statistically significant reduction in the luteolin and xanthohumol cohorts, while the genus *Clostridia UCG-014* was increased in the same cohorts (3.07% in luteolin cohort and 5.10% in xanthohumol cohort vs. 1.57% in the PBS cohort). The genus *Lachnospiraceae NK4A136* group (family *Lachnospiraceae*) showed a statistically significant increase in the luteolin cohort (15.48% vs. 6.19% in the PBS cohort), while the genus *Blautia*, belonging to the same family, was significantly increased in both the apigenin (0.30%) and the luteolin (0.37% vs. 0.07% in the PBS cohort) cohorts. The genus *Ruminococcus* (family *Ruminococcaceae*) showed a statistically significant reduction in the luteolin cohort (1.48% vs. 6.19% in the PBS cohort). The genus *Alistipes* (family *Rikenellaceae*) showed a high statistically significant decrease in the xanthohumol cohort (0.39% vs. 7.66% in the PBS cohort). The observed reduction in the phylum *Pseudomonadota* in the luteolin and xanthohumol cohorts was due to reductions in the genera *Parasutterella* and *Escherichia-Shigella*. The genus *Adlercreutzia* (family *Eggerthellaceae*) showed a statistically significant reduction in the luteolin cohort (0.17% vs. 0.4% in the PBS cohort). Conversely, the genus *Enterorhabdus* showed a high increase in the luteolin and xanthohumol cohorts (0.55% and 0.46% vs. 0.17% in the PBS cohort) (Table 3, Supplementary Figure S4).

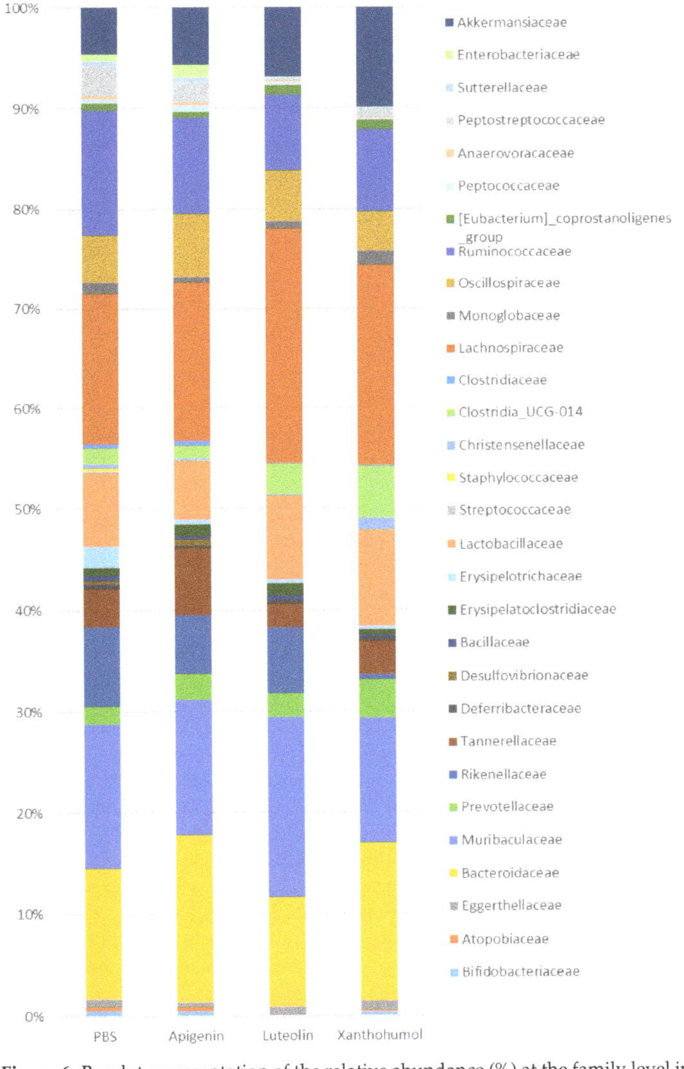

Figure 6. Barplot representation of the relative abundance (%) at the family level in the gut microbiota populations between the animals induced with UC across the four experimental cohorts. See Table 2 for a detailed list of statistically significant differences between the PBS control animals and each of the flavonoid treatments in the UC-induced animals.

Table 2. Average relative abundance (%) at the family level in the gut microbiota between animals induced with UC across the four experimental cohorts. Significant differences in the comparison between the control PBS cohort and each flavonoid treatment cohort are depicted. API: apigenin; LUT: luteolin; XAN: xanthohumol. Asterisks indicate statistically significant differences (* $p < 0.05$; ** $p < 0.005$; *** $p < 0.0005$; **** $p < 0.0001$).

Family	PBS	Apigenin	Luteolin	Xanthohumol	PBS vs. API	PBS vs. LUT	PBS vs. XAN
Bifidobacteriaceae	0.52	0.43	0.03	0.35		**	
Atopobiaceae	0.42	0.43	0.007	0.18		**	
Eggerthellaceae	0.68	0.43	0.79	0.89			
Bacteroidaceae	12.66	16.23	10.59	15.25			
Muribaculaceae	14.15	13.16	17.56	12.16			
Prevotellaceae	1.73	2.48	2.38	3.83			*
Rikenellaceae	7.76	5.71	6.39	0.46			****
Tannerellaceae	3.74	6.51	2.33	3.22			
Deferribacteraceae	0.46	0.29	0.33	0.35			
Desulfovibrionaceae	0.36	0.53	0.07	0.03			*
Bacillaceae	0.45	0.29	0.37	0.13			
Erysipelatoclostridiaceae	0.79	1.09	1.20	0.57			
Erysipelotrichaceae	2.08	0.53	0.38	0.36	*	**	**
Lactobacillaceae	7.08	5.77	8.04	9.31			
Streptococcaceae	0.16	0.02	0.003	0.02	*	***	*
Staphylococcaceae	0.35	0.05	0	0.003	**	***	****
Christensenellaceae	0.36	0.10	0.08	1.07	*	*	
Clostridia UCG-014	1.57	1.30	3.07	5.10		**	****
Clostridiaceae	0.4	0.43	0.11	0.12		***	**
Lachnospiraceae	14.96	15.65	23.25	19.75		**	
Monoglobaceae	1.08	0.59	0.71	1.41			
Oscillospiraceae	4.63	6.21	4.92	3.89			
Ruminococcaceae	12.28	9.39	7.38	7.93			
Eubacterium	0.75	0.58	0.93	0.97			
Peptococcaceae	0.39	0.63	0.35	0.17			**
Anaerovoracaceae	0.33	0.29	0.13	0.11		**	***
Peptostreptococcaceae	2.88	1.96	0.21	0.86		****	*
Sutterellaceae	0.54	0.45	0.11	0.08		**	****
Enterobacteriaceae	0.68	1.30	0.05	0.03		**	**
Akkermansiaceae	4.56	5.57	6.80	9.77			*

At the species level (Table 3, Supplementary Figure S5), an uncharacterized bacterium from the genus *Blautia* showed a statistically significant increase in all three flavonoid treatment cohorts (0.28% in the apigenin cohort, 0.33% in the luteolin cohort and 0.15% in the xanthohumol cohort vs. 0.01% in the PBS cohort). The observed changes in the family *Bifidobacteriaceae* are mainly associated with the species *Bifidobacterium animalis,* which was highly reduced in the luteolin cohort. The two species *Bacteoides dorei* (1.57% in xanthohumol vs. 0.87% in PBS cohort) and *Bacteroides thetaiotaomicron* (2.32% in xanthohumol vs. 1.31% in PBS cohort) showed an increase in the xanthohumol cohort, and *B. dorei* also increased in the apigenin cohort (1.59%), but decreased in the luteolin cohort (0.36%).

Regarding the comparisons between the three flavonoid treatment cohorts, the luteolin and xanthohumol cohorts only showed differences in the genera *Alistipes*, *Peptococcus*, and *Romboutsia* (Supplementary Figures S3 and S4). The apigenin cohort shared the same differences as the PBS cohort in comparison to the other two cohorts: *Bifidobacteriaceae, Atopobiaceae, Rickenellaceae, Desulfovibrionaceae, Staphylococcaceae, Clostridia UCG-014, Clostridiaceae, Peptococcaceae, Peptostreptococcacee, Suterellaceae,* and *Enterobacteriaceae*. The apigenin cohort showed main differences compared to the luteolin cohort regarding the genera *Parabacteroides* and *Christenellaceae R7* group and compared to the xanthohumol cohort in the genus *Christenellaceae R7* group.

Table 3. Average relative abundance (%) at the genus and species levels in the gut microbiota between animals induced with UC across the four experimental cohorts. Significant differences in the comparison between the control PBS cohort and each flavonoid treatment cohort are depicted. Asterisks indicate statistically significant differences (* $p < 0.05$; ** $p < 0.005$; *** $p < 0.0005$; **** $p < 0.0001$). API: apigenin; LUT: luteolin; XAN: xanthohumol.

Genus Species	PBS	Apigenin	Luteolin	Xanthohumol	PBS vs. API	PBS vs. LUT	PBS vs. XAN
Bifidobacterium	0.52	0.43	0.03	0.35		**	
B. animalis	0.41	0.41	0.03	0.28		*	
Adlercreutzia	0.40	0.22	0.17	0.24		*	
Enterorhabdus	0.17	0.15	0.55	0.46		****	***
Bacteroides	12.66	16.23	10.59	15.25			
B. dorei	0.87	1.59	0.36	1.57	*	**	*
B. thetaiotaomicron	1.31	2.09	1.12	2.32			*
Alistipes	7.66	5.67	6.37	0.39			****
Bilophila	0.36	0.53	0.07	0.03			*
Turicibacter	0.85	0.35	0.10	0.22		****	*
Streptococcus	0.15	0.02	0.003	0.016		**	*
Staphylococcus	0.35	0.05	0	0.003	***	***	****
Clostridia_UCG-014	1.57	1.30	3.07	5.10		**	****
Clostridium sensu stricto 1	0.40	0.43	0.11	0.12		***	**
Blautia	0.07	0.30	0.37	0.19	*	*	
Uncharacterized	0.01	0.28	0.33	0.15	**	**	*
Lachnospiraceae NK4A136	6.19	7.32	15.48	12.27		*	
Ruminococcus	6.19	3.47	1.48	2.93		**	
Peptococacceae uncultured	0.36	0.53	0.31	0.17			***
Romboutsia	2.88	1.96	0.21	0.86		****	*
Parasutterella	0.54	0.45	0.11	0.08		**	****
Escherichia-Shigella	0.68	1.30	0.05	0.03		**	**
Akkermansia	4.56	5.57	6.80	9.77			*
A. muciniphila	4.56	5.57	6.80	9.77			*

3. Discussion

UC is characterized by chronic inflammation and ulcers in the colon mucosa. This study evaluated the anti-inflammatory potential of IP injection of three different flavonoids (apigenin, luteolin, and xanthohumol) in the treatment of UC, identifying any direct effect, as well as potential changes in the gut microbiota composition that may explain or correlate with any beneficial effect in the UC pathophysiology observed (cytokines levels, mucosa alterations, etc.). These three compounds have previously demonstrated in vitro antitumor activity against HT-29, HCT116, and T84 human colon cancer cell lines [19]. Furthermore, oral administration of these flavonoids to murine models showed in vivo efficacy in the treatment of IBD. Assessments included a reduction in anti-inflammatory markers, mitigation of colon injuries, restoration of intestinal barrier integrity, and downregulation of immune pathways. An evaluation of the effects on gut microbiota modulation was conducted solely for apigenin and luteolin [1,20–23]. Conversely, oral administration of quercetin failed in the co-treatment enhancement of the anticancer effect compared to IP administration, further emphasizing the potential advantageous features of IP injection. Additionally, increased levels of quercetin were observed in tumor tissues following IP administration [29]. Using the IP administration route in this work, these nutraceuticals avoid GI first-pass and they are first transferred to the mesentery capillaries; from there, they reach the portal vein towards the liver. From the liver, they are secreted via the bile duct to the small intestine, where they finally reach the gut microbiota, modulating the taxa present in the lumen and generating metabolism products (derived from the corresponding flavonoid skeleton), which may act at the colonocyte level or enter the portal circulation again [25].

In this study, a total of forty rats were utilized as a model for UC induced by DSS, and the progression of this condition was assessed across various parameters. The rats were divided into four cohorts, with three cohorts receiving IP injections of flavonoids and the fourth cohort serving as a control (receiving only PBS injections). Following UC induction, each flavonoid treatment was compared to the PBS control cohort in terms of efficacy. Biomarkers from body, tissue, and plasma samples were analyzed, alongside metataxonomic data of the gut microbiota. No adverse effects on the animals' health were observed following flavonoid treatments, as indicated by biomarker comparisons between healthy control animals (those sentinels not subjected to UC induction) across the different cohorts.

Among the three evaluated flavonoids, the present study has shown a superior effectiveness of luteolin in mitigating UC across all measured parameters. In this regard, a total absence of colon mucosa ulceration (Figure 2d) and a high reduction in the DAI index (Figure 2a,c) could be assessed after luteolin administration, in accordance with previous observations by other authors using oral administration [1]. The reduction in the DAI index in the luteolin cohort could be mainly attributed to an improvement in the stool consistency score (Figure 2b) since no significant differences were observed in body weight gain between the different cohorts (Supplementary Figure S1a). Although not statistically significant, it was notable how the decrease in the body weight on days 9 to 11, due to the UC stage, was markedly more pronounced in the PBS and apigenin cohorts, indicating a more acute manifestation of UC in these two cohorts (Figure 1).

The reductions observed in the pro-inflammatory IL-6 and IL-1β cytokines (Figure 3a,b) may be attributed to either the direct systemic anti-inflammatory properties of the flavonoids or the generation of bioactive metabolites resulting from the microbial metabolism of these flavonoids in the gut. Most known microbial gut metabolites derived from the flavones apigenin and luteolin comprise 3-(4′-hydroxyphenyl) propionic acid and 3-(3′,4′-dihydroxyphenyl) propionic acid, respectively, generated by the cleavage of C-ring, with phloroglucinol being released in both cases [24,26,27,30]. In the case of xanthohumol, the resulting gut metabolite is the potent phytoestrogen 8-prenylnaringenin [26]. These gut microbiota metabolites have been described as antioxidant, anticancer, and antimicrobial bioactives [31], as well as anti-inflammatory, and are able to inhibit the secretion of pro-inflammatory cytokines (TNF-α, IL-1β, and IL-6) [32]. In our particular assay, xanthohumol, which has been described as being degraded by some microorganisms present in the human gut microbiota [24,26], caused a reduction in IL-6 circulating levels but had no significant effect on the IL-1β plasma levels in the analyzed rats, while apigenin and luteolin, which are metabolized in a different way, seemed to have similar effects over both cytokines in our animal model for UC (Figure 3a,b).

According to the literature, the gut microbiota plays a critical role in the pathogenesis of UC [33–35]. Dysbiotic gut microbiota was reported to be necessary for inflammation, and thus, this is associated with the development and progression of this IBD [12,36]. In this regard, a metataxonomics analysis of the cecal content was performed and the gut microbiota composition was determined through 16S ribosomal RNA sequencing. As a result, some interesting changes in the gut microbiota composition have emerged here as the most significant ones among all the comparisons performed. Based on these findings, we can affirm that IP administration enables flavonoids to reach the colon lumen through mesenteric absorption into the portal vein, their subsequent transfer to the liver, and eventual secretion into the intestinal lumen via the bile duct, thereby exerting an effect on the intestinal luminal ecosystem.

Thus, while richness and evenness were similar in all the cohorts of this study (Supplementary Figure S2), interesting differences were observed in terms of beta diversity of gut microbiota composition between the different groups (Figure 4). These beta diversity analyses suggest that the PBS and apigenin cohorts showed a similar gut microbiota community structure, while the same cohorts are barely related to the luteolin and xanthohumol cohorts, which in turn, seemed to share a resemblance (Figure 4b). This observation also became apparent when a heatmap clustering of the samples based on the abundance of genera was depicted (Supplementary Figure S6). Subsequently, a comprehensive analysis of the microbiota reaffirmed this finding (Figure 6, Tables 2 and 3).

Major statistically significant differences were observed in the comparison of UC animals' gut microbiota after luteolin and/or xanthohumol administration compared to the control PBS cohort. These differences included reductions and increases in pro-inflammatory and anti-inflammatory bacteria, respectively. The genera *Turicibacter*, *Streptococcus*, *Staphylococcus*, *Clostridium sensu stricto 1*, *Romboutsia*, *Parasutterella*, and *Escherichia-Shigella* [37–40] were reduced both in the luteolin and the xanthohumol cohorts (Table 3, Supplementary Figure S4). The *Parasutterella* and *Escherichia-Shigella* genera (phylum *Pseudomonadota*) are considered part of an unstable gut microbial community, and they are associated with the genesis and development of IBD, as well as with chronic intestinal inflammation [37,41,42]. Conversely, the genera *Clostridia UCG-014* and *Enterorhabdus* [26,43] were increased in these luteolin and xanthohumol cohorts (Table 3, Supplementary Figure S4). In particular, the genus *Enterorhabdus* belongs to the polyphenol-degrading family *Eggerthellaceae* (phylum *Actinomycetota*), which is associated with health status and reported to potentiate the production of the previously mentioned bioactive phenolic metabolites [26,44,45].

Changes associated only with the xanthohumol cohort were also observed, including a reduction in the family *Peptococcaceae* and the genera *Alistipes* and *Bilophila* (Tables 2 and 3) [28,46,47]. The genus *Alistipes*, highly reduced in the UC animals from the xanthohumol cohort, has been correlated with the development of dysbiosis and disease [48]. The genus *Bilophila* has been described to have a potential role in chronic inflammation [49], and it has been found to be reduced in animal models treated orally with anthocyanins [28]. In the phylum *Bacteroidota*, statistically significant increases were observed in the family *Prevotellaceae* in the xanthohumol cohort (Table 2, Supplementary Figure S3) [39,50], as well as in the species *Bacteroides dorei* and *Bacteroides thetaiotaomicron*, both being species reported with anti-inflammatory activity (Table 3, Supplementary Figure S5) [51,52]. The species *Akkermansia muciniphila* (phylum *Verrucomicrobiota*) is mainly considered a potential second-generation probiotic for the treatment of intestinal microbiome-associated diseases [53]. In this study, *A. muciniphila* was increased in the UC animals of the xanthohumol cohort (Table 3, Supplementary Figure S5). This species is a mucin-degrading bacteria reported to release monosaccharides, amino acids, and SCFAs into the environment, stimulating beneficial intestinal bacteria metabolic functions and contributing to the alleviation of microbial dysbiosis due to IBD [54–56].

Meanwhile, changes exclusively associated with the luteolin cohort included increases in the genera *Blautia* and *Lachnospiraceae NK4A136* group, both belonging to the family *Lachnospiraceae*, and reductions in the genus *Ruminococcus* (Table 3) [37,39,57,58]. The family *Lachnospiraceae* (Table 2) is one of the main butyrate-producing bacteria, and it has been associated with the control of gut inflammatory processes via the reduction of the expression of pro-inflammatory cytokines, the conversion of primary to secondary bile acids, and the resistance against the colonization of intestinal pathogens [59–62]. The increase in the genus *Lachnospiraceae NK4A136* group has been previously related to inflammation alleviation after polyphenol administration [63]. Interestingly, an uncharacterized bacterium was increased in all three flavonoid treatment cohorts (Table 3, Supplementary Figure S5). This species belongs to the genus *Blautia*, a commensal SCFAs-producing bacteria, with a role in maintaining the environmental balance in the intestine and in preventing inflammation by upregulating intestinal regulatory T_{reg} cells [57]. Although a significant reduction was observed in the common probiotic species *Bifidobacterium animalis* in the UC animals of the luteolin cohort in comparison with the PBS cohort ones, this was the result of a maintained abundance of this species in all animals (healthy controls and UC-induced) in the luteolin cohort (Table 3, Supplementary Figure S5). Consistent with this observation, the literature has previously reported a bacteriostatic effect of luteolin over lactic acid bacteria and *B. animalis* [64].

Collectively, although numerous studies have reported the effectiveness of apigenin in the amelioration of IBD [20,21], our investigations demonstrate that, in the context of this UC animal model, the outcomes associated with apigenin treatment closely resemble those of the control cohort administered with PBS. This similarity extends across multiple domains, including physiological parameters, tissue biomarkers, and microbiota analyses, but not in the case of cytokines, where apigenin administration actually causes significant statistical differences in plasma levels of IL-1β and IL-6 (a direct effect of flavonoids IP injections).

In contrast, luteolin, and to a lesser extent xanthohumol, emerged as promising therapeutic agents for the prevention and treatment of UC disease when administered via IP injection. Both luteolin and xanthohumol exhibited an ability to modulate the composition of the gut microbiota community, favoring the enrichment of anti-inflammatory taxa, capable of producing advantageous compounds, like SCFAs or phenolic metabolites derived from flavonoids, while concurrently reducing the presence of pro-inflammatory taxa within the intestinal lumen. They could potentially elicit their beneficial effects towards UC directly, and their effects can be mediated through the bioactive metabolites produced by the gut microbiota during flavonoid metabolism.

A primary contributing factor for the observed lower efficacy of apigenin, with a profile more akin to the PBS cohort, compared to luteolin and xanthohumol, in terms of modulation of gut ecosystem, appeared to be its reduced bioavailability following IP administration. This assertion was substantiated by the conspicuous presence of apigenin precipitate granules dispersed throughout the mesentery after euthanasia (Supplementary Figure S7). These deposits of apigenin suggested that, in contrast to luteolin and xanthohumol, apigenin experienced limited translocation into the mesentery capillaries, thus preventing its ultimate entry into the gut lumen. This disparity in translocation could likely be attributed to a lower solubility of apigenin, stemming from its chemical structure, wherein it possesses only one hydroxyl group on ring C, in contrast to luteolin, which possesses two.

The primary limitation of this study was mainly related to the timing of euthanasia, which occurred several days after the peak of symptoms in the UC challenge, during the animals' recovery period (day 12). This timing may have avoided the detection of peak concentrations of pro-inflammatory cytokines in plasma and MPO in colon mucosa. Conversely, the study of gut microbiota populations during the symptomatic peak of UC (days 7 to 8) would have rendered a substantially altered composition of the gut microbiota and structural changes in the colon mucosa. During this period, the potential therapeutic effects of the administered flavonoids may have been masked due to factors such as reduced food intake by the animals and significant alterations in the digestive tract caused by the presence of blood.

The main contributions of this work include the detection of a lower bioavailability in the case of apigenin when administered intraperitoneally, probably due to its lower hydrophilicity. Also, there was a strong reduction in IL-6 and IL-1β cytokines exerted by the flavonoids and a reduction in the UC-associated weight loss (a symptom caused by the DSS treatment during pathology onset) in the case of luteolin and xanthohumol. At the tissular level, the strong reduction caused by luteolin in the DAI index and colon ulceration is remarkable. Finally, numerous and very significant changes were exerted by luteolin and xanthohumol regarding colon microbiota modulation (a strong increase in some anti-inflammatory populations and a strong reduction in some pro-inflammatory and commensal taxa), an aspect that is entirely new for this pathology in the case of xanthohumol.

4. Materials and Methods

4.1. Drugs and Chemicals

Dextran Sodium Sulfate (DSS, 40,000 g/mol) was purchased from VWR Chemicals (Madrid, Spain). Apigenin and luteolin were provided by Fluorochem (Dublin, Ireland). Xanthohumol was purified following a modified procedure described previously [65]. The same batch of spent hops, stored in high-density polyethylene (HDPE) industrial barrels and closed under a nitrogen atmosphere, was used. The modification involved only the initial extract preparation step as it was fully completed at the Department of Food Chemistry and Biocatalysis, Wrocław University of Environmental and Life Sciences laboratories. Eighteen kilograms of spent hops were extracted with 90 L of acetone in 0.2 kg:1.4 L batches, each made in a 2-L Erlenmeyer flask shaken for 3 h on a rotary shaker (120 rpm). The formed pulp was vacuum-filtered on Whatman filter paper no. 4 and concentrated using a laboratory rotary evaporator. The combined extracts were further subjected to polyphenol precipitation and Sephadex LH-20 column chromatography steps, resulting in 20.233 g of Xanthohumol (>98% purity by HPLC). The flavonoids were resuspended for injection in Tween 80 and phosphate-buffered saline (PBS) at a stock concentration of 5 mg/mL for apigenin and luteolin and 2.5 mg/mL for xanthohumol.

4.2. Animals

A total of 40 five-week-old male Fischer 344 rats (*Rattus norvegicus*), were maintained in the Animal Facilities at the University of Oviedo (authorized facility No. ES330440003591). All animal experiments were approved by the Ethics Committee of the Principality of Asturias (authorization code PROAE 29/2021).

The rats were provided by Charles River (Lyon, France). Sterile drinking water and standard pelleted feed (Teklad Irradiated Global 14% Protein Rodent Maintenance Diet) (Envigo, Gannat, France) were provided ad libitum. The animals were housed in a room under controlled temperature (21 °C) and humidity and 12 h light/darkness cycles.

4.3. Experimental Design

After being acclimated for 1 week, the animals were randomized into four cohorts of 10 animals each, according to the administered treatment. These animals were intraperitoneally injected with PBS (cohort 1, negative control) or the corresponding flavonoid treatment (10 mg/kg of body weight): apigenin (cohort 2), luteolin (cohort 3), or xanthohumol (cohort 4). The IP injections were administered daily since the start of the experiment and for a period of eleven days.

Eight rats from each experimental group underwent UC induction through the addition of 3% DSS to the autoclaved drinking water, which was administered ad libitum from the initiation of the experiment and for one week. The remaining two animals from each group served as absolute healthy control animals and were not subjected to UC induction. These absolute control animals were utilized as sentinels to assess the effects of the treatment compounds (flavonoids) on healthy animals and to detect any possible side (negative) effects.

On day twelve, the animals were anesthetized (with isofluorane) (Zoetis, Madrid, Spain) and sacrificed (via pneumothorax). Two rats, designated as rat number 6 within the luteolin-treated cohort and rat number 3 within the xanthohumol-treated cohort, necessitated euthanasia on days eight and nine of the experiment, respectively, due to severe weight loss conditions.

The parameters that were monitored daily during the whole experiment included body weight, food and water intake, stool consistency, and rectal bleeding.

4.4. Tissue Sample Collection

After euthanasia, 2 mL of blood was extracted via heart puncture from each animal and centrifuged at 3000 rpm for 15 min and then, the plasmas were frozen. The small intestine was removed, and then, the hyperplastic Peyer's patches were macroscopically quantified. The caecums were immediately weighed after extraction using a precision scale and then frozen at −20 °C. Finally, the colons were removed, opened longitudinally, and washed with PBS (VWR, Madrid, Spain) in order to macroscopically assess the length and ulceration status. Also, proximal and distant samples from colons were collected and frozen at −80 °C.

4.5. Histological Studies

The hyperplastic Peyer's patches were counted along each small intestine. Their number in the experimental animals was compared with respect to the hyperplastic Peyer's patches of the 2 absolute control animals from each cohort (animals 9 and 10).

The percentage of colon length reduction in the UC animals was compared with respect to the colon length of the 2 control (healthy) animals from each cohort.

Regarding the macroscopic score assessment of UC, this parameter was quantified as follows: 0—no symptoms; 1—local hyperemia but no ulceration; 2—ulceration without hyperemia; 3—ulceration and inflammation in only one site; 4—two or more ulceration and inflammation sites; 5—ulceration bigger than 2 cm; values 6 to 1—one score point per each 1 cm of extra ulceration [66].

4.6. Assessment of the Disease Activity Index (DAI)

In order to quantify the clinical evolution of the UC, the Disease Activity Index (DAI) [66] was used. This index is a numerical disease activity measurement comprising the sum of two parameters: changes in growth rate (0: more than 5% body weight gain; 1: less than 5% body weight gain and less than 5% body weight loss; 2: 5 to 10% body weight loss; 3: 10 to 20% body weight loss; 4: more than 20% body weight loss), and stool consistency score (0: normal feces; 1: loose stool; 2: watery diarrhea; 3: slimy diarrhea with little blood; 4: severe watery diarrhea with blood).

4.7. Pro-Inflammatory Cytokine Analysis in Plasma

IL-1β and IL-6 tests were performed in plasma samples, using commercial Rat IL-1β and IL-6 ELISA Kits (Diaclone, Besançon, France) and following the manufacturer's instructions.

4.8. Myeloperoxidase (MPO) Assays

A 0.5-cm longitudinal section from each distal colon was excised, and this pro-inflammatory enzyme was quantified following a published protocol [66].

4.9. 16S rRNA Sequencing and Gut Microbiota Analysis

A metagenomics analysis of the stool samples, obtained from the caecums, was also performed. For this, the Pathogen Detection Protocol from the E.Z.N.A.® Stool DNA Kit (VWR, Madrid, Spain) was used. Caecums were thawed on ice for 30 min, and then, 200 mg of feces from a middle section of each caecum were placed in a 25-mL tube to continue with the extraction protocol. Finally, 200 uL of genomic DNA were obtained and quantified using a BioPhotometer® (Eppendorf, Madrid, Spain). The total DNA samples were frozen at −20 °C in order to be subsequently analyzed via the amplification and sequencing of the variable regions V3 and V4 of the 16S ribosomal RNA gene using Illumina MiSeq (Microomics Systems, Barcelona, Spain). Amplification was performed after 25 PCR cycles. A negative control of the DNA extraction as well as a positive Mock Community control were included to ensure quality control. These studies allowed us to describe and quantify microbial alpha and beta diversities, as well as the taxonomic profiles from phylum to species levels.

4.10. Microbiota Analysis

Phylotype data were used to calculate the alpha diversity metrics in order to analyze the diversity of microbial communities. Alpha diversity analysis was used to measure the community richness (observed Operating Taxonomic Units or OTUs), defined as the number of different phylotypes present in a community. Alpha diversity was also used to measure the community evenness, given as the Pielou's evenness index, which quantifies how equal the community is numerically, taking into account the number and the abundance of phylotypes in a community. The Chao1 (species richness), Simpson (level of biodiversity), and Shannon (species diversity) indices were also calculated.

Phylotype and phylogenetic data were used to calculate the beta diversity metrics in order to assess the microbial community's structure. A principal coordinate analysis (PCoA) was performed, based on unweighted Unifrac distance, a phylogenetic qualitative measure, in order to detect differences in beta diversity.

4.11. Statistical Methods

For the metataxonomics analysis, alpha diversity comparisons were performed using a linear model with the appropriate distribution (negative binomial model for Chao1, beta regression for Simpson, and linear model for Shannon). Beta diversity distance matrices were used to calculate PCoA and to make ordination plots using the R software package version 4.2.0. The significance of the groups present in the community structure was tested using the Permanova test. The differential relative abundance of taxa was tested using a linear model based on the negative binomial distribution and ANOVA. Biodiversity R version 2.14-1, PMCMRplus version 1.9.4, RVAideMemoire version 0.9-8, and Vegan version 2.5-6 packages were used for the different statistical analyses carried out.

For the rest of the comparisons, outliers have been identified and excluded from the statistical analyses. The normality of the different variables was tested using Shapiro–Wilk's test. In light of these results, the data were then expressed as the mean value ± standard error of the mean (S.E.M.), and parametric methods were used for statistical analyses. Differences between cohorts were tested by a one-way ANOVA (analysis of variance). When the quantitative data were not normal, the non-parametric Kruskal–Wallis test was used. The graphic representation was carried out using GraphPad Prism software (version 9, GraphPad Software, San Diego, CA, USA). In each case, a p-value < 0.05 was considered statistically significant (* $p < 0.05$; ** $p < 0.005$; *** $p < 0.0005$; **** $p < 0.0001$).

Supplementary Materials: The following supporting information can be downloaded at: https://www.mdpi.com/article/10.3390/ijms25063236/s1.

Author Contributions: Investigation, P.M.-C., Á.P.-V., and S.Y.; Resources, S.S., J.P., and E.H.; Supervision, F.L. and C.J.V.; Writing—original draft preparation, P.M.-C.; Writing—review and editing, F.L.; Funding acquisition, F.L. All authors have read and agreed to the published version of the manuscript.

Funding: This research was funded by Principado de Asturias (Spain) through the program "Ayu-das a organismos públicos para apoyar las actividades de I+D+I de sus grupos de investigación" (grant AYUD/2021/51347), "Programa Severo Ochoa de Ayudas Predoctorales para la investigación y docencia" from Principado de Asturias (grant PA-20-PF-BP19-058 to P.M.-C. and grant PA-21-PF-BP20-150 to Á.P.-V.), the research project PID2021-127812OB-I00 from MICINN (Spanish Ministry of Science and Innovation), and the European Union's Horizon 2020 Research and Innovation Program under Grant Agreement no. 814650 for the project SynBio4Flav.

Institutional Review Board Statement: The animal study protocol was approved by the Ethics Committee of the Principality of Asturias (protocol code PROAE 29/2021, 29 October 2021).

Informed Consent Statement: Not applicable.

Data Availability Statement: Publicly available datasets (metagenome sequences) were analyzed in this study. These data can be found in the NCBI SRA database with accession number PRJNA1068569.

Acknowledgments: The authors wish to thank the University of Oviedo Animal Facility Unit (Servicios Científico-Técnicos, SCTs).

Conflicts of Interest: The authors declare no conflicts of interest. The funders had no role in the design of this study, the collection, analyses, or interpretation of the data, the writing of the manuscript, or the decision to publish the results.

References

1. Li, B.; Du, P.; Du, Y.; Zhao, D.; Cai, Y.; Yang, Q.; Guo, Z. Luteolin Alleviates Inflammation and Modulates Gut Microbiota in Ulcerative Colitis Rats. *Life Sci.* **2021**, *269*, 119008. [CrossRef] [PubMed]
2. De Souza, H.S.P.; Fiocchi, C.; Iliopoulos, D. The IBD Interactome: An Integrated View of Aetiology, Pathogenesis and Therapy. *Nat. Rev. Gastroenterol. Hepatol.* **2017**, *14*, 739–749. [CrossRef]

3. Carding, S.; Verbeke, K.; Vipond, D.T.; Corfe, B.M.; Owen, L.J. Dysbiosis of the Gut Microbiota in Disease. *Microb. Ecol. Health Dis.* **2015**, *26*, 26191. [CrossRef] [PubMed]
4. Mosca, A.; Leclerc, M.; Hugot, J.P. Gut Microbiota Diversity and Human Diseases: Should We Reintroduce Key Predators in Our Ecosystem? *Front. Microbiol.* **2016**, *7*, 455. [CrossRef] [PubMed]
5. Oligschlaeger, Y.; Yadati, T.; Houben, T.; Condello Oliván, C.M.; Shiri-Sverdlov, R. Inflammatory Bowel Disease: A Stressed "Gut/Feeling". *Cells* **2019**, *8*, 659. [CrossRef] [PubMed]
6. Hirano, A.; Umeno, J.; Okamoto, Y.; Shibata, H.; Ogura, Y.; Moriyama, T.; Torisu, T.; Fujioka, S.; Fuyuno, Y.; Kawarabayasi, Y.; et al. Comparison of the Microbial Community Structure between Inflamed and Non-inflamed Sites in Patients with Ulcerative Colitis. *J. Gastroenterol. Hepatol.* **2018**, *33*, 1590–1597. [CrossRef] [PubMed]
7. Nishino, K.; Nishida, A.; Inoue, R.; Kawada, Y.; Ohno, M.; Sakai, S.; Inatomi, O.; Bamba, S.; Sugimoto, M.; Kawahara, M.; et al. Analysis of Endoscopic Brush Samples Identified Mucosa-Associated Dysbiosis in Inflammatory Bowel Disease. *J. Gastroenterol.* **2018**, *53*, 95–106. [CrossRef] [PubMed]
8. Atreya, R.; Neurath, M.F. Involvement of IL-6 in the Pathogenesis of Inflammatory Bowel Disease and Colon Cancer. *Clin. Rev. Allergy Immunol.* **2005**, *28*, 187–196. [CrossRef]
9. Eastaff-Leung, N.; Mabarrack, N.; Barbour, A.; Cummins, A.; Barry, S. Foxp3+ Regulatory T Cells, Th17 Effector Cells, and Cytokine Environment in Inflammatory Bowel Disease. *J. Clin. Immunol.* **2010**, *30*, 80–89. [CrossRef]
10. Liso, M.; Verna, G.; Cavalcanti, E.; De Santis, S.; Armentano, R.; Tafaro, A.; Lippolis, A.; Campiglia, P.; Gasbarrini, A.; Mastronardi, M.; et al. Interleukin 1β Blockade Reduces Intestinal Inflammation in a Murine Model of Tumor Necrosis Factor–Independent Ulcerative Colitis. *Cell Mol. Gastroenterol. Hepatol.* **2022**, *14*, 151–171. [CrossRef]
11. Shahini, A.; Shahini, A. Role of Interleukin-6-Mediated Inflammation in the Pathogenesis of Inflammatory Bowel Disease: Focus on the Available Therapeutic Approaches and Gut Microbiome. *J. Cell Commun. Signal* **2023**, *17*, 55–74. [CrossRef]
12. Lee, M.; Chang, E.B. Inflammatory Bowel Diseases (IBD) and the Microbiome—Searching the Crime Scene for Clues. *Gastroenterology* **2021**, *160*, 524–537. [CrossRef]
13. Himmel, M.E.; Yao, Y.; Orban, P.C.; Steiner, T.S.; Levings, M.K. Regulatory T-cell Therapy for Inflammatory Bowel Disease: More Questions than Answers. *Immunology* **2012**, *136*, 115–122. [CrossRef] [PubMed]
14. Jochum, L.; Stecher, B. Label. or Concept—What Is a Pathobiont? *Trends Microbiol.* **2020**, *28*, 789–792. [CrossRef] [PubMed]
15. Ahmad, M.S.; Krishnan, S.; Ramakrishna, B.S.; Mathan, M.; Pulimood, A.B.; Murthy, S.N. Butyrate and Glucose Metabolism by Colonocytes in Experimental Colitis in Mice. *Gut* **2000**, *46*, 493–499. [CrossRef]
16. Belzer, C.; Chia, L.W.; Aalvink, S.; Chamlagain, B.; Piironen, V.; Knol, J.; de Vos, W.M. Microbial Metabolic Networks at the Mucus Layer Lead to Diet-Independent Butyrate and Vitamin B 12 Production by Intestinal Symbionts. *mBio* **2017**, *8*, e00770-17. [CrossRef] [PubMed]
17. Macfarlane, S.; Macfarlane, G.T. Regulation of Short-Chain Fatty Acid Production. *Proc. Nutr. Soc.* **2003**, *62*, 67–72. [CrossRef] [PubMed]
18. Zhao, H.; Cheng, N.; Zhou, W.; Chen, S.; Wang, Q.; Gao, H.; Xue, X.; Wu, L.; Cao, W. Honey Polyphenols Ameliorate DSS-Induced Ulcerative Colitis via Modulating Gut Microbiota in Rats. *Mol. Nutr. Food Res.* **2019**, *63*, e1900638. [CrossRef] [PubMed]
19. Fernández, J.; Silván, B.; Entrialgo-Cadierno, R.; Villar, C.J.; Capasso, R.; Uranga, J.A.; Lombó, F.; Abalo, R. Antiproliferative and Palliative Activity of Flavonoids in Colorectal Cancer. *Biomed. Pharmacother.* **2021**, *143*, 112241. [CrossRef] [PubMed]
20. Fu, R.; Wang, L.; Meng, Y.; Xue, W.; Liang, J.; Peng, Z.; Meng, J.; Zhang, M. Apigenin Remodels the Gut Microbiota to Ameliorate Ulcerative Colitis. *Front. Nutr.* **2022**, *9*, 1062961. [CrossRef]
21. Ai, X.-Y.; Qin, Y.; Liu, H.-J.; Cui, Z.-H.; Li, M.; Yang, J.-H.; Zhong, W.-L.; Liu, Y.-R.; Chen, S.; Sun, T.; et al. Apigenin Inhibits Colonic Inflammation and Tumorigenesis by Suppressing STAT3-NF-KB Signaling. *Oncotarget* **2017**, *8*, 100216–100226. [CrossRef] [PubMed]
22. Xue, L.; Jin, X.; Ji, T.; Li, R.; Zhuge, X.; Xu, F.; Quan, Z.; Tong, H.; Yu, W. Luteolin Ameliorates DSS-Induced Colitis in Mice via Suppressing Macrophage Activation and Chemotaxis. *Int. Immunopharmacol.* **2023**, *124*, 110996. [CrossRef] [PubMed]
23. Cho, J.-M.; Yun, S.-M.; Choi, Y.-H.; Heo, J.; Kim, N.-J.; Kim, S.-H.; Kim, E.-H. Xanthohumol Prevents Dextran Sulfate Sodium-Induced Colitis via Inhibition of IKKβ/NF-KB Signaling in Mice. *Oncotarget* **2018**, *9*, 866–880. [CrossRef]
24. Braune, A.; Gutschow, M.; Engst, W.; Blaut, M. Degradation of Quercetin and Luteolin by Eubacterium Ramulus. *Appl. Environ. Microbiol.* **2001**, *67*, 5558–5567. [CrossRef] [PubMed]
25. Al Shoyaib, A.; Archie, S.R.; Karamyan, V.T. Intraperitoneal Route of Drug Administration: Should It Be Used in Experimental Animal Studies? *Pharm. Res.* **2020**, *37*, 12. [CrossRef] [PubMed]
26. Makarewicz, M.; Drożdż, I.; Tarko, T.; Duda-Chodak, A. The Interactions between Polyphenols and Microorganisms, Especially Gut Microbiota. *Antioxidants* **2021**, *10*, 188. [CrossRef] [PubMed]
27. Ozdal, T.; Sela, D.A.; Xiao, J.; Boyacioglu, D.; Chen, F.; Capanoglu, E. The Reciprocal Interactions between Polyphenols and Gut Microbiota and Effects on Bioaccessibility. *Nutrients* **2016**, *8*, 78. [CrossRef] [PubMed]
28. Fernández, J.; García, L.; Monte, J.; Villar, C.; Lombó, F. Functional Anthocyanin-Rich Sausages Diminish Colorectal Cancer in an Animal Model and Reduce Pro-Inflammatory Bacteria in the Intestinal Microbiota. *Genes* **2018**, *9*, 133. [CrossRef]
29. Chan, S.-T.; Lin, Y.-C.; Chuang, C.-H.; Shiau, R.-J.; Liao, J.-W.; Yeh, S.-L. Oral and Intraperitoneal Administration of Quercetin Decreased Lymphocyte DNA Damage and Plasma Lipid Peroxidation Induced by TSA In Vivo. *Biomed. Res. Int.* **2014**, *2014*, 580626. [CrossRef]

30. Marín, L.; Miguélez, E.M.; Villar, C.J.; Lombó, F. Bioavailability of Dietary Polyphenols and Gut Microbiota Metabolism: Antimicrobial Properties. *Biomed. Res. Int.* **2015**, *2015*, 905215. [CrossRef]
31. Zieniuk, B. Dihydrocaffeic Acid—Is It the Less Known but Equally Valuable Phenolic Acid? *Biomolecules* **2023**, *13*, 859. [CrossRef]
32. Li, H.; Christman, L.M.; Li, R.; Gu, L. Synergic Interactions between Polyphenols and Gut Microbiota in Mitigating Inflammatory Bowel Diseases. *Food Funct.* **2020**, *11*, 4878–4891. [CrossRef]
33. Nishihara, Y.; Ogino, H.; Tanaka, M.; Ihara, E.; Fukaura, K.; Nishioka, K.; Chinen, T.; Tanaka, Y.; Nakayama, J.; Kang, D.; et al. Mucosa-Associated Gut Microbiota Reflects Clinical Course of Ulcerative Colitis. *Sci. Rep.* **2021**, *11*, 13743. [CrossRef]
34. Qiu, P.; Ishimoto, T.; Fu, L.; Zhang, J.; Zhang, Z.; Liu, Y. The Gut Microbiota in Inflammatory Bowel Disease. *Front. Cell Infect. Microbiol.* **2022**, *12*, 733992. [CrossRef]
35. Cui, L.; Guan, X.; Ding, W.; Luo, Y.; Wang, W.; Bu, W.; Song, J.; Tan, X.; Sun, E.; Ning, Q.; et al. *Scutellaria baicalensis* Georgi Polysaccharide Ameliorates DSS-Induced Ulcerative Colitis by Improving Intestinal Barrier Function and Modulating Gut Microbiota. *Int. J. Biol. Macromol.* **2021**, *166*, 1035–1045. [CrossRef]
36. Gophna, U.; Sommerfeld, K.; Gophna, S.; Doolittle, W.F.; Veldhuyzen van Zanten, S.J.O. Differences between Tissue-Associated Intestinal Microfloras of Patients with Crohn's Disease and Ulcerative Colitis. *J. Clin. Microbiol.* **2006**, *44*, 4136–4141. [CrossRef] [PubMed]
37. Shang, L.; Liu, H.; Yu, H.; Chen, M.; Yang, T.; Zeng, X.; Qiao, S. Core Altered Microorganisms in Colitis Mouse Model: A Comprehensive Time-Point and Fecal Microbiota Transplantation Analysis. *Antibiotics* **2021**, *10*, 643. [CrossRef]
38. Bajer, L.; Kverka, M.; Kostovcik, M.; Macinga, P.; Dvorak, J.; Stehlikova, Z.; Brezina, J.; Wohl, P.; Spicak, J.; Drastich, P. Distinct Gut Microbiota Profiles in Patients with Primary Sclerosing Cholangitis and Ulcerative Colitis. *World J. Gastroenterol.* **2017**, *23*, 4548. [CrossRef]
39. Zou, J.; Shen, Y.; Chen, M.; Zhang, Z.; Xiao, S.; Liu, C.; Wan, Y.; Yang, L.; Jiang, S.; Shang, E.; et al. Lizhong Decoction Ameliorates Ulcerative Colitis in Mice via Modulating Gut Microbiota and Its Metabolites. *Appl. Microbiol. Biotechnol.* **2020**, *104*, 5999–6012. [CrossRef] [PubMed]
40. Wang, H.; Zhang, M.; Wen, X.; He, L.; Zhang, M.; Zhang, J.; Yang, X. Cepharanthine Ameliorates Dextran Sulphate Sodium-induced Colitis through Modulating Gut Microbiota. *Microb. Biotechnol.* **2022**, *15*, 2208–2222. [CrossRef]
41. Chen, Y.; Wu, H.; Wu, S.; Lu, N.; Wang, Y.; Liu, H.; Dong, L.; Liu, T.; Shen, X. Parasutterella, in Association with Irritable Bowel Syndrome and Intestinal Chronic Inflammation. *J. Gastroenterol. Hepatol.* **2018**, *33*, 1844–1852. [CrossRef]
42. Li, W.; Cheng, F.; Zhang, J.; Li, C.; Yu, D.; Simayijiang, H.; Liu, H.; Li, S.; Yan, J. Changes in Gut Microbiota and Metabolites in Papillary Thyroid Carcinoma Patients Following Radioactive Iodine Therapy. *Int. J. Gen. Med.* **2023**, *16*, 4453–4464. [CrossRef]
43. Yu, Z.; Xiaojia, L.; Wei, Z.; Jian, Z.; Aiting, W.; Jing, W.; Lin, Y.; Bangwei, C.; Dan, Y. Baicalin Circumvents Anti-PD-1 Resistance by Regulating the Gut Microbiota Metabolite Short-Chain Fatty Acids. *Pharmacol. Res.* **2024**, *199*, 107033. [CrossRef]
44. Rodríguez-Daza, M.-C.; Roquim, M.; Dudonné, S.; Pilon, G.; Levy, E.; Marette, A.; Roy, D.; Desjardins, Y. Berry Polyphenols and Fibers Modulate Distinct Microbial Metabolic Functions and Gut Microbiota Enterotype-Like Clustering in Obese Mice. *Front. Microbiol.* **2020**, *11*, 2032. [CrossRef]
45. Taladrid, D.; Zorraquín-Peña, I.; Molinero, N.; Silva, M.; Manceñido, N.; Pajares, R.; Bartolomé, B.; Moreno-Arribas, M.V. Polyphenols and Ulcerative Colitis: An Exploratory Study of the Effects of Red Wine Consumption on Gut and Oral Microbiome in Active-Phase Patients. *Mol. Nutr. Food Res.* **2022**, *66*, e2101073. [CrossRef]
46. Chen, L.; Li, X.; Gu, Q. *Chimonanthus salicifolius* Extract Alleviates DSS-induced Colitis and Regulates Gut Microbiota in Mice. *Food Sci. Nutr.* **2023**, *11*, 3019–3030. [CrossRef]
47. Rowan, F.E.; Docherty, N.G.; Coffey, J.C.; O'Connell, P.R. Sulphate-Reducing Bacteria and Hydrogen Sulphide in the Aetiology of Ulcerative Colitis. *Br. J. Surg.* **2009**, *96*, 151–158. [CrossRef]
48. Parker, B.J.; Wearsch, P.A.; Veloo, A.C.M.; Rodriguez-Palacios, A. The Genus Alistipes: Gut Bacteria with Emerging Implications to Inflammation, Cancer, and Mental Health. *Front. Immunol.* **2020**, *11*, 906. [CrossRef]
49. Feng, Z.; Long, W.; Hao, B.; Ding, D.; Ma, X.; Zhao, L.; Pang, X. A Human Stool-Derived *Bilophila wadsworthia* Strain Caused Systemic Inflammation in Specific-Pathogen-Free Mice. *Gut Pathog.* **2017**, *9*, 59. [CrossRef]
50. Wang, M.; Zhou, B.; Cong, W.; Zhang, M.; Li, Z.; Li, Y.; Liang, S.; Chen, K.; Yang, D.; Wu, Z. Amelioration of AOM/DSS-Induced Murine Colitis-Associated Cancer by Evodiamine Intervention Is Primarily Associated with Gut Microbiota-Metabolism-Inflammatory Signaling Axis. *Front. Pharmacol.* **2021**, *12*, 797605. [CrossRef]
51. Delday, M.; Mulder, I.; Logan, E.T.; Grant, G. Bacteroides Thetaiotaomicron Ameliorates Colon Inflammation in Preclinical Models of Crohn's Disease. *Inflamm. Bowel Dis.* **2019**, *25*, 85–96. [CrossRef]
52. Yoshida, N.; Emoto, T.; Yamashita, T.; Watanabe, H.; Hayashi, T.; Tabata, T.; Hoshi, N.; Hatano, N.; Ozawa, G.; Sasaki, N.; et al. *Bacteroides vulgatus* and *Bacteroides dorei* Reduce Gut Microbial Lipopolysaccharide Production and Inhibit Atherosclerosis. *Circulation* **2018**, *138*, 2486–2498. [CrossRef]
53. He, X.; Bai, Y.; Zhou, H.; Wu, K. *Akkermansia muciniphila* Alters Gut Microbiota and Immune System to Improve Cardiovascular Diseases in Murine Model. *Front. Microbiol.* **2022**, *13*, 6920. [CrossRef]
54. Zheng, M.; Han, R.; Yuan, Y.; Xing, Y.; Zhang, W.; Sun, Z.; Liu, Y.; Li, J.; Mao, T. The Role of *Akkermansia muciniphila* in Inflammatory Bowel Disease: Current Knowledge and Perspectives. *Front. Immunol.* **2023**, *13*, 1089600. [CrossRef]
55. Xue, L.; Zhao, Y.; Wang, H.; Li, Z.; Wu, T.; Liu, R.; Sui, W.; Zhang, M. The Effects of Live and Pasteurized *Akkermansia muciniphila* on DSS-Induced Ulcerative Colitis, Gut Microbiota, and Metabolomics in Mice. *Food Funct.* **2023**, *14*, 4632–4646. [CrossRef]

56. Xu, Y.; Wang, N.; Tan, H.-Y.; Li, S.; Zhang, C.; Feng, Y. Function of *Akkermansia muciniphila* in Obesity: Interactions with Lipid Metabolism, Immune Response and Gut Systems. *Front. Microbiol.* **2020**, *11*, 219. [CrossRef]
57. Liu, X.; Mao, B.; Gu, J.; Wu, J.; Cui, S.; Wang, G.; Zhao, J.; Zhang, H.; Chen, W. Blautia—A New Functional Genus with Potential Probiotic Properties? *Gut Microbes* **2021**, *13*, 1875796. [CrossRef]
58. Hall, A.B.; Yassour, M.; Sauk, J.; Garner, A.; Jiang, X.; Arthur, T.; Lagoudas, G.K.; Vatanen, T.; Fornelos, N.; Wilson, R.; et al. A Novel Ruminococcus Gnavus Clade Enriched in Inflammatory Bowel Disease Patients. *Genome Med.* **2017**, *9*, 103. [CrossRef]
59. Ćesić, D.; Lugović Mihić, L.; Ozretić, P.; Lojkić, I.; Buljan, M.; Šitum, M.; Zovak, M.; Vidović, D.; Mijić, A.; Galić, N.; et al. Association of Gut Lachnospiraceae and Chronic Spontaneous Urticaria. *Life* **2023**, *13*, 1280. [CrossRef]
60. Lobionda, S.; Sittipo, P.; Kwon, H.Y.; Lee, Y.K. The Role of Gut Microbiota in Intestinal Inflammation with Respect to Diet and Extrinsic Stressors. *Microorganisms* **2019**, *7*, 271. [CrossRef]
61. Vacca, M.; Celano, G.; Calabrese, F.M.; Portincasa, P.; Gobbetti, M.; De Angelis, M. The Controversial Role of Human Gut *Lachnospiraceae*. *Microorganisms* **2020**, *8*, 573. [CrossRef]
62. Zhang, K.; Zhu, L.; Zhong, Y.; Xu, L.; Lang, C.; Chen, J.; Yan, F.; Li, J.; Qiu, J.; Chen, Y.; et al. Prodrug Integrated Envelope on Probiotics to Enhance Target Therapy for Ulcerative Colitis. *Adv. Sci.* **2023**, *10*, 2205422. [CrossRef]
63. Xia, T.; Duan, W.; Zhang, Z.; Li, S.; Zhao, Y.; Geng, B.; Zheng, Y.; Yu, J.; Wang, M. Polyphenol-Rich Vinegar Extract Regulates Intestinal Microbiota and Immunity and Prevents Alcohol-Induced Inflammation in Mice. *Food Res. Int.* **2021**, *140*, 110064. [CrossRef]
64. Puupponen-Pimia, R.; Nohynek, L.; Meier, C.; Kahkonen, M.; Heinonen, M.; Hopia, A.; Oksman-Caldentey, K.-M. Antimicrobial Properties of Phenolic Compounds from Berries. *J. Appl. Microbiol.* **2001**, *90*, 494–507. [CrossRef]
65. Nowak, B.; Poźniak, B.; Popłoński, J.; Bobak, Ł.; Matuszewska, A.; Kwiatkowska, J.; Dziewiszek, W.; Huszcza, E.; Szeląg, A. Pharmacokinetics of Xanthohumol in Rats of Both Sexes after Oral and Intravenous Administration of Pure Xanthohumol and Prenylflavonoid Extract. *Adv. Clin. Exp. Med.* **2020**, *29*, 1101–1109. [CrossRef]
66. Fernández, J.; de la Fuente, V.G.; García, M.T.F.; Sánchez, J.G.; Redondo, B.I.; Villar, C.J.; Lombó, F. A Diet Based on Cured Acorn-Fed Ham with Oleic Acid Content Promotes Anti-Inflammatory Gut Microbiota and Prevents Ulcerative Colitis in an Animal Model. *Lipids Health Dis.* **2020**, *19*, 28. [CrossRef]

International Journal of
Molecular Sciences

MDPI

Article

Molecular Characterization of a Restriction Endonuclease PsaI from *Pseudomonas anguilliseptica* KM9 and Sequence Analysis of the PsaI R-M System

Beata Furmanek-Blaszk *, Iwona Mruk and Marian Sektas

Department of Microbiology, Faculty of Biology, University of Gdansk, Wita Stwosza 59, 80-308 Gdansk, Poland; iwona.mruk@ug.edu.pl (I.M.); marian.sektas@ug.edu.pl (M.S.)
* Correspondence: beata.furmanek-blaszk@ug.edu.pl

Abstract: A restriction enzyme PsaI, an isoschizomer of the type II restriction endonuclease HindIII, has been purified to homogeneity from Gram-negative bacilli *Pseudomonas anguilliseptica* KM9 found in a wastewater treatment plant in Poland. Experimental data revealed that R.PsaI is highly active in the presence of Co^{2+}, Mg^{2+}, and Zn^{2+} and reached a maximal level of activity between 2.5 and 10 mM while its activity was significantly decreased in the presence of Ca^{2+}, Fe^{2+}, Mn^{2+}, and Ni^{2+}. Moreover, we found that the purified R.PsaI did not require NaCl for enzyme activity. Restriction cleavage analysis followed by sequencing confirmed 5′-AAGCTT-3′ as the recognition site. The genes for restriction–modification system PsaI were identified and characterized. Downstream of the psaIM gene, we noticed an ORF that shares extensive similarity with recombinase family protein specifically involved in genome rearrangements. Sequence analysis revealed that the PsaI R-M gene complex showed striking nucleotide sequence similarity (>98%) with the genes of the PanI R-M system from a *P. anguilliseptica* MatS1 strain identified in a soil sample from Sri Lanka.

Keywords: restriction–modification systems; type II restriction endonuclease PsaI; isoschizomer HindIII

1. Introduction

The two enzymes that make up type II restriction–modification (R-M) systems are a methyltransferase, which transfers a methyl group to a specific nucleotide to prevent cleavage, and a restriction endonuclease, which identifies and cleaves a particular DNA sequence at a specific site. Four groups of restriction endonucleases have been identified based on the DNA recognition sequence, cleavage position, cofactor requirements, and protein subunit composition [1]. Type II enzymes have been studied extensively because they play a fundamental role in gene analysis and cloning work. Most of them require divalent metal ions to catalyze the cleavage of the phosphodiester bond. Although the natural cofactor for all type II restriction endonucleases is Mg^{2+}, they can utilize a variety of divalent cations for in vitro DNA cleavage reaction, including Ca^{2+}, Cd^{2+}, Co^{2+}, Fe^{2+}, Mn^{2+}, Ni^{2+}, or Zn^{2+} [2,3].

Many R-M enzymes have been isolated from various microorganisms over the years, but the necessity of restriction endonucleases with unique specificities for molecular cloning has largely motivated the search. Because type II restriction enzymes can identify and cut DNA sequences and produce a predictable cleavage pattern, they are most frequently utilized in molecular biology applications. This makes them essential for recombinant DNA technology [4].

Among the over 4000 biochemically or genetically characterized restriction enzymes found to date, 98% belong to type II [4]. This high percentage could result from their great demand and ease of detection, although it might not accurately represent the distribution of R-M systems. Type II restriction endonucleases generally recognize the palindromic sequence in DNA and cleave within or near the recognition sequence, producing DNA fragments of defined sizes. These properties have led to the screening of diverse bacterial species for new restriction activities. Up to now, numerous target sites have been recognized by various restriction endonucleases, many of which have new specificities that have not yet been found.

In this work, we investigated samples taken from water pollution control plants at each stage of sewage treatment. Bacterial communities found in wastewater treatment plants (WWTPs) can be quite varied, consisting of hundreds of species [5]. In this heterogeneous ecosystem, bacteria-dominant microorganisms belong to the phyla Firmicutes, Proteobacteria, and Bacteroidetes [6]. *Pseudomonas* spp. are able to thrive in diverse ecological niches, including highly nutritious environments, such as sewage [7].

High abundance, density, heterogeneity, activity, and complex interplay within activated sludge bioreactors, like those in WWTPs, would appear to be linked to increased rates of gene flow, including both horizontal and vertical transfer of genes. Taking into account the vast amount of bacteria passing through WWTPs every day, it is likely that many of them could serve as a potential source of various enzymes. Sludge is also increasingly being viewed and treated as a source of valuable new products originating from bacteria.

To date, many restriction endonucleases have been found in various bacteria, some of which are isoschizomers used as tools in molecular biology. In this work, we investigated samples from the wastewater treatment plant "Gdynia-Debogorze", and the isolated bacteria were screened for the presence of restriction endonucleases. One of the isolates found in the activated sludge, identified as *P. anguilliseptica* KM9, produced the restriction endonuclease PsaI. In this manuscript, we describe the isolation, purification, and characterization of the enzyme and analyze its properties. R.PsaI belongs to type II restriction endonucleases, and it is an isoschizomer of the HindIII enzyme from *Haemophiluus influenzae*. The *P. anguilliseptica* KM9 strain seems to be a promising source of R.PsaI possessing HindIII-type restriction activity.

2. Results and Discussion

2.1. Detection of Restriction Endonucleases Activity in Bacterial Isolates

WWTPs are the habitat of many types of microorganisms, which makes them interesting considering their potential to produce important enzymes for biotechnological exploration. Furthermore, the production of restriction endonucleases in this particular environment provides bacterial cells with powerful protection against a broad spectrum of phages. Years of screening various microbial sources have yielded hundreds of restriction enzymes that recognize and cleave specific DNA sequences. Knowing that R-M systems have been detected in bacteria from all ecological niches and taxonomic groups, including WWTPs, we aimed to screen this microbiome resource and search for novel microbial restriction endonucleases. In this work, a good method to screen for type II restriction endonucleases consisted of incubation on cell extracts with known DNA substrates [8].

By employing the quick screening method to analyze the restriction activities in the crude cell extracts, restriction endonucleases were found in at least 18 of the 320 isolates. Only one strain, identified as *P. anguilliseptica* KM9, showed restriction activity, which gave a clear, sharp, banding pattern on lambda DNA similar to that observed with lambda genome digested with HindIII restriction endonuclease. Hence, it was selected for further study (Figure S1). According to the suggested nomenclature rules, this enzyme was named PsaI [9].

2.2. Enzyme Purification

Because our laboratory has been involved in studies to characterize R-M systems, we report here the isolation and characterization of R.PsaI. The purification process involved four chromatographic steps. The crude cell extract contained a set of contaminating proteins, and most of these were eliminated through phosphocellulose chromatography. The strategy used for purifying R.PsaI is summarized in Table 1.

Table 1. Purification of restriction endonuclease R.PsaI from *Pseudomonas anguilliseptica* KM9.

R.PsaI Purification Step	Total Protein (mg)	Total Activity (U × 10^3)	Specific Activity (U mg^{-1})	Yield (%)	Purification (Fold)
Cell free extract	652	10,000	15.54 × 10^3	1	100
Phosphocellulose	48	2920	60.83 × 10^3	4	29
Hydroxylapatite	13.5	1280	94.81 × 10^3	6	12.8
CM Sephadex	1.35	450	333.3 × 10^3	22	4.5

The final enzyme preparation was free of non-specific nucleases. From 10 g of bacteria, we were able to obtain 1.35 mg of homogenous enzyme preparation, with an overall yield of 4.5%. R.PsaI was analyzed through SDS-PAGE to determine its purity and denaturated molecular mass. According to Coomassie blue-stained gels, the enzyme was found to be at least 95% pure. Relative to the standards of a known Mr, a value of Mr = 35,000 ± 100 for R.PsaI was calculated (Figure S2). Enzyme stability was tested after different periods of storage at −20 °C; the purified enzyme did not lose activity after storage for 12 months in buffer with 50% glycerol.

The simplicity of cultivation, presence of the single restriction endonuclease, and high level of its production make *P. anguilliseptica* KM9 a promising producer of PsaI restriction endonuclease isoschizomeric to R.HindIII for use in experimental practice in industry. To date (14 May 2025), 269 isoschizomers of HindIII restriction endonucleases have been found in many bacterial genera. However, only one has been found in the genus *Pseudomonas* [4]. They are reported in the Restriction Enzyme database (http://rebase.neb.com) (accessed on 14 May 2025), although only R.HindIII is commercially available from fourteen suppliers.

2.3. Optimal Conditions for Restriction Activity

The effect of several factors on R.PsaI activity was investigated. To identify the optimal buffer and temperature, the enzyme was tested for its ability to digest 150 ng of lambda DNA substrate with one unit of the enzyme in one hour. The purified enzyme was active over a temperature range of 30–52 °C (Figure 1).

When the cleavage activity of R.PsaI was tested in different commercial buffers using lambda DNA as a substrate, the formation of a characteristic cleavage pattern was observed in all of the tested buffers. Upon digestion of lambda DNA with R.PsaI for 24 h, no traces of unspecific cleavage products were noted.

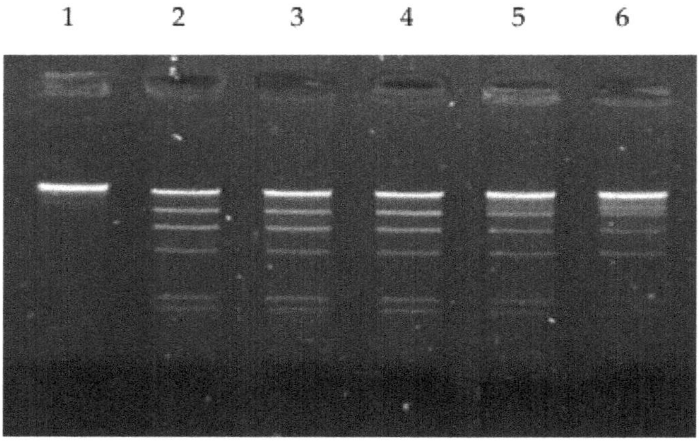

Figure 1. Determination of optimum temperature of the purified PsaI restriction enzyme activity. Digestion of lambda DNA was carried out in NEB2 buffer for 1 h at lane 2 at 30 °C, lane 3 at 37 °C, lane 4 at 42 °C, lane 5 at 52 °C, and lane 6 at 60 °C. Lane 1—undigested lambda DNA.

2.4. Effect of Divalent Metal Ions and Ionic Strength on R.PsaI Activity

The activity of most enzymes is easily affected by environmental factors, such as metal ions that are usually found in wastewater. Therefore, one of the objectives of this study included investigating the effect of various concentrations of some metals, referred to as trace elements, on the activity of the restriction enzyme we isolated from *P. anguilliseptica* KM9. Type II restriction endonucleases require certain cofactors to digest DNA. Almost all restriction endonucleases of the PD-(D/E)XK family need divalent cations—usually Mg^{2+}, which is a natural cofactor for the majority of these enzymes—for DNA cleavage [10]. Although the physiological metal ion for the restriction endonucleases is magnesium, they can utilize a variety of divalent cations for in vitro cleavage reaction [11]. Firstly, to identify whether other metal ions can promote endonucleolytic activity of R.PsaI, we performed an array of reactions using different divalent cations. To determine R.PsaI activity in the presence of various divalent cations, such as Ca^{2+}, Co^{2+}, Fe^{2+}, Mg^{2+}, Mn^{2+}, Ni^{2+}, or Zn^{2+}, we used plasmid pUC18 DNA containing a single recognition site as a substrate and then analyzed the products through gel electrophoresis. The results showed an absolute requirement for divalent cations for R.PsaI cleavage activity, preferably Co^{2+} and Mg^{2+} (Figure 2). Four metal ions (calcium, iron, manganese, and nickel) failed to support cleavage, whereas zinc displayed only a small reduction in cleavage activity.

Many type II restriction endonucleases can use Mn^{2+} in place of Mg^{2+} as a cofactor, but only a few can use a broader range of divalent cations instead [3].

To investigate more precisely the requirements of R.PsaI, we carried out DNA cleavage experiments at various divalent cation concentrations ranging from 2.5 to 10 mM in the presence of plasmid pUC18 DNA containing a single recognition site as a substrate, and then we analyzed the products through gel electrophoresis (Figure S3). R.PsaI shows a qualitatively similar Co^{2+}, Mg^{2+}, and Zn^{2+} concentration dependence. In general, the maximum level of activity was observed in the presence of Mg^{2+}, Co^{2+}, and Zn^{2+} at concentrations between 5 and 10 mM. We also noticed that R.PsaI showed a slight preference for higher Co^{2+} concentrations compared to Mg^{2+} or Zn^{2+} (between 2.5 and 10 mM) to maintain the same cleavage efficiency.

Our results suggest that R.PsaI has a metal-binding site that can accept varied divalent metal ions, modulating the catalytic activity. In industrial wastewaters, various metal ions are present, but only some of them are relevant in the environmental context. The ability to utilize a wide range of metal cofactors for DNA cleavage may be of great importance to the biological function of R.PsaI residing in *P. anguilliseptica* KM9. When fighting with other bacteria for their environment and phages that have the potential to infect the host cell, this characteristic may help the bacteria break down invasive foreign genomes.

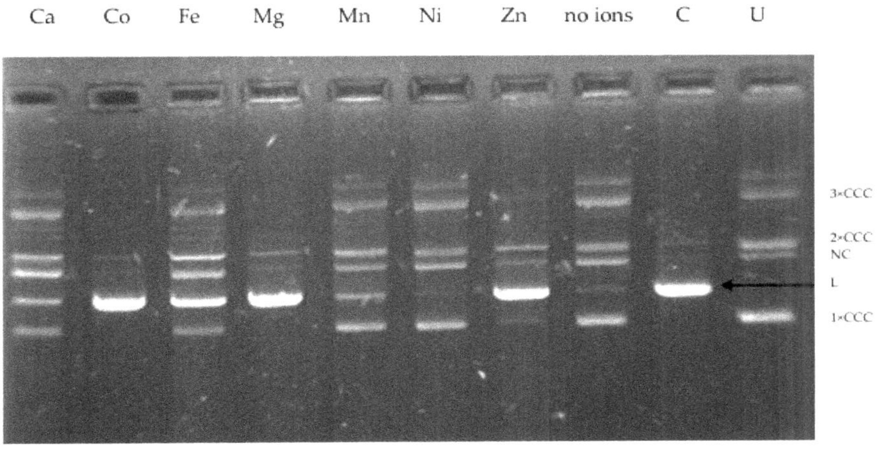

Figure 2. Effect of divalent metal ions on the activity of R.PsaI. Cleavage reactions of pUC18 DNA were carried out in the presence of 10 mM of Ca^{2+}, Co^{2+}, Fe^{2+}, Mg^{2+}, Mn^{2+}, Ni^{2+}, and Zn^{2+} or without any metal ion. C—control lane containing pUC18 digested with R.PsaI in NEB2 buffer; U—undigested pUC18 DNA, where an arrow indicates the linear form of pUC18 DNA (2686 bp); 3 × CCC—supercoiled plasmid trimer; 2 × CCC—supercoiled plasmid dimer; 1 × CCC—supercoiled plasmid monomer; NC—nicked circular plasmid DNA; L—linear form of pUC18.

Restriction enzymes show different specificity in their response to ionic strength. The effect of ionic strength on R.PsaI activity was determined by varying the NaCl concentrations from 0 to 250 mM. The enzyme is remarkably tolerant to high concentrations of NaCl, being active up to 200 mM NaCl. The activity decreased at a concentration higher than 200 mM (partial DNA digestion) (Figure 3). Moreover, in the absence of NaCl, lambda DNA cleavage was comparable to that observed in the presence of a wide range of sodium chloride concentrations.

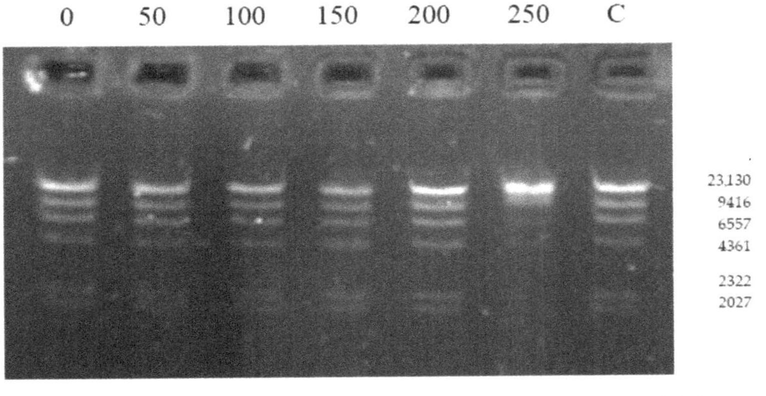

Figure 3. Effect of the NaCl on R.PsaI activity. Cleavage reactions of lambda DNA were carried out in the presence of different NaCl concentrations (0, 50, 100, 150, 200, 250 mM). C—control lane containing lambda DNA incubated with R.PsaI in NEB2 buffer.

R.PsaI retains activity with or without NaCl and thus differs from other restriction endonucleases, which require salt to remain active. For that reason, R.PsaI could be a valuable tool for DNA manipulation as it is active at a broad range of salt concentrations.

2.5. Determination of the R.PsaI Cleavage Site

To determine the recognition sequence and cleavage site of R.PsaI, plasmid pGEM3Zf(+) was digested with purified R.PsaI. Then, the linear form of DNA was sequenced using GS5 and GS6 primers (Table S1) (Figure 4). The sequencing of linear plasmid DNA revealed the recognition sequence 5′-AAGCTT-3′, where the cleavage occurs between two adenines.

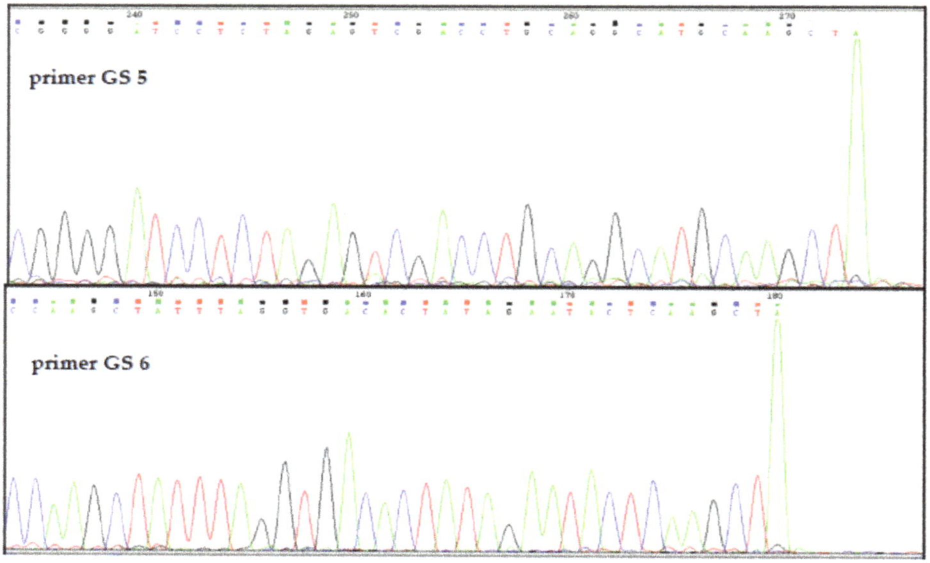

Figure 4. Run-off sequencing to determine the R.PsaI cleavage site in pGEM3Zf(+). The drop in the peak signal indicates where the DNA polymerase runs off of the template at the nicked site. Modified DNA polymerase adds an additional adenine (A) at the end of the extension product.

2.6. Genomic DNA Modification Status

The results described above prompted us to examine the pattern of DNA methylation. The presence of cognate methyltransferase activity was assessed based on its ability to protect genomic DNA from cleavage by the PsaI restriction enzyme. *P. anguilliseptica* KM9 displayed methylation protection against its endogenous restriction enzyme (Figure 5), as well as isoschizomers of R.HindIII, while digestion by restriction endonucleases with different specificities resulted in DNA cleavage and subsequent degradation (Figure 5).

The data revealed the existence of bacterial methyltransferase activity, which suggested the presence of a complete type II restriction–modification system in *P. anguilliseptica* KM9 cells. It should be noted that this protection against R.PsaI only indicates the possibility, but does not prove, that M.PsaI has identical specificity.

This finding prompted us to screen the *P. anguilliseptica* KM9 DNA sequence for the presence of a gene coding for cognate PsaI methyltransferase. Studies with R-M systems revealed that restriction endonucleases show little primary sequence similarity among themselves, whereas methyltransferases share substantial sequence homology and could be identified based on primary sequence data [12]. Knowing that the greatest similarity between methyltransferases of R-M systems is restricted to the two most conserved motifs

(I and IV) [13,14], we designed primers for amplification of a potential methyltransferase PsaI gene containing conserved motifs I and IV. The forward primer (isometDIPY) (Table S1) was complementary to the S-adenosylmethionine binding motif (I), whereas the reverse primer (isometR) (Table S1) was complementary to the methylation catalytic motif (IV). To determine the partial nucleotide sequence of the gene coding for M.PsaI, genomic DNA of *P. anguilliseptica* KM9 was subjected to PCR amplification of the selected methyltransferase coding DNA fragment. This pair of primers produced an amplicon at a size of around 700 bp, showing sequence similarity to site-specific DNA methyltransferases. The presence and distribution of highly conserved amino acid sequence motifs indicated that M.PsaI is a member of the N^6-adenine methylases family. Because genes encoding R-M systems are usually closely linked, we decided to analyze DNA fragments located upstream and downstream of the identified *psaIM* gene fragment.

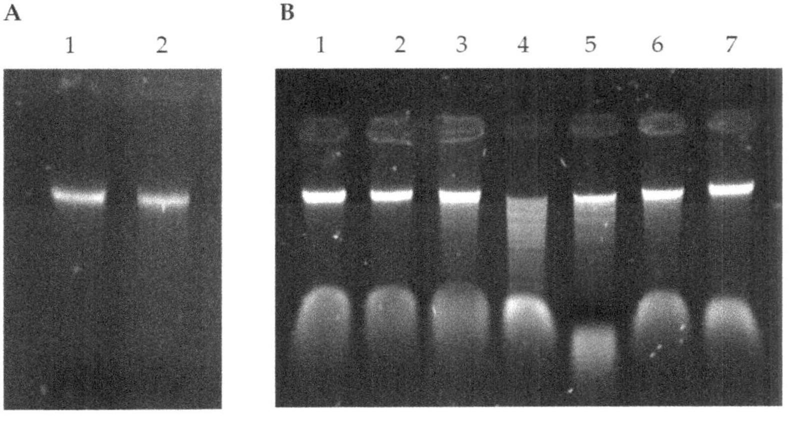

Figure 5. Digestion of the genomic DNA of the *Pseudomonas anguilliseptica* KM9 strain with restriction endonucleases. Panel (**A**): R.PsaI—lane 2; lane 1 shows undigested DNA as a control; panel (**B**): R.HindIII—lane 2, R.EcoRI—lane 3, R.EcoRV—lane 4, R.KpnI—lane 5, R.BamHI—lane 6, and R.EcoVIII—lane 7; lane 1 shows undigested DNA as a control.

For this, the iPCR approach, which allows for amplification of the unknown sequences adjacent to the known DNA fragment, was employed. For this purpose, we designed primers isometF and isometinv1 (Table S1) oriented in opposite directions. Nucleotide sequencing of the obtained PCR product revealed the presence of a DNA fragment at a size of around 800 pb showing sequence similarity to type II restriction endonuclease HindIII. Using this technique, we determined the nucleotide sequence of a 1.8-kilobase DNA stretch containing two incomplete open reading frames showing significant similarity to genes coding for restriction and modification enzymes isospecific to R-M system HindIII. However, for an unknown reason, using the iPCR technique, we were unable to amplify either the 5′ or 3′ termini of *psaIR* and *psaIM* genes.

In 2022, an article describing the isolation and partial characterization of a new restriction endonuclease PanI, isoschizomer of R.HindIII, from the *P. anguilliseptica* MatS1 strain was published [15]. We noticed that the *panIR* gene shared a striking similarity to the known portion of the *psaIR* gene we identified in the *P. anguilliseptica* KM9 strain. Knowing that genes that code for restriction enzymes nearly always occur next to the genes that code for the corresponding methyltransferases, we analyzed sequences surrounding the *panIR* gene and noticed that the methyltransferase gene *panIM* lies immediately adjacent to the *panIR* gene. This finding indicated the presence of a complete R-M system in the *P. anguilliseptica* MatS1 strain.

Then, in order to identify the lacking sequences corresponding to the 5′ and 3′ ends of *psaIR* and *psaIM* genes, respectively, we designed primers based on published sequences of genes encoding the R-M system PanI [15] (resPsastart and metPsaend) (Table S1). Nucleotide sequencing of the obtained PCR products uncovered the presence in *P. anguilliseptica* KM9 of an R-M gene complex almost identical to R-M system PanI. Protein alignment results have shown that R.PsaI and M.PsaI share about 98% identity with their bacterial counterparts of *P. anguilliseptica* MatS1.

2.7. PsaI R-M System Analysis

In this work, we also defined R-M system PsaI residing in the *P. anguilliseptica* KM9 strain. The PsaI R-M system is organized in a head to tail orientation often found in many R-M systems. Among a group of R-M systems isospecific to HindIII, only the prototype R-M system from *Haemophilus influenzae* Rd has a similar genetic organization (Figure 6) [16].

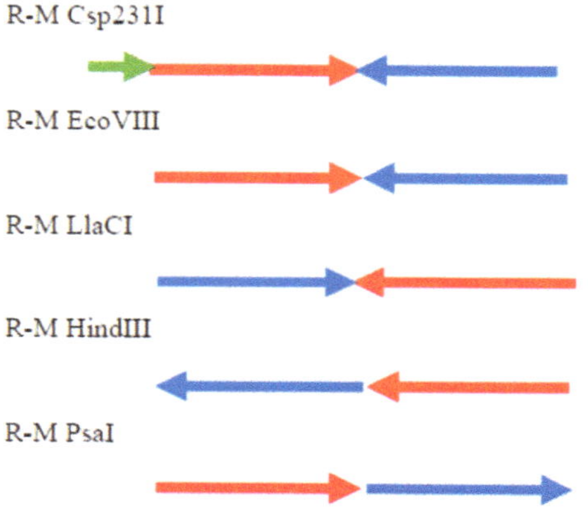

Figure 6. The organization of gene clusters encoding HindIII-like R-M systems. Arrows represent the direction of translation and the relative sizes of open reading frames. The control protein is highlighted in green, restriction endonucleases in red, and methyltransferases in blue.

The G + C content of the genes encoding the complete PsaI R-M system is 43.73%, which is lower than the overall guanine–cytosine composition of *P. anguilliseptica* genomes (60%) [17]. The presence of R-M genes with different GC content suggests that they could have been integrated into bacterial genomes through horizontal gene transfer. The restriction endonuclease and modification methylase genes lie adjacent to each other and are oriented transcriptionally in a sequential manner. The start codon of M.PsaI is placed just before the stop codon of R.PsaI. The first gene of the PsaI R-M system, *psaIR*, encodes a protein with a calculated molecular mass of 34,633 Da, which consisted of 304 amino acids. Database searches showed that this protein, apart from a remarkable similarity to R.PanI (98%) [15], also shares significant homology to the HindIII family type II restriction endonucleases found in many bacteria. The greatest level of similarity (≥70%) was observed between R.PsaI and proteins from *Marinospirillum* sp. (78.62%), *Cylindrospermopsis raciborskii* (75.66%), *Candidatus Nitrotoga fabula* (76.32%), and *Negativicutes bacterium* (73.36%) (for details, see [15]). The alignment of the R.PsaI amino acids sequence with well-studied isospecific restriction endonucleases of R-M systems, such as R.EcoVIII, R.LlaCI, R.Csp231I, and R.HindIII, revealed homology of 56.62%, 28.57%, 21.89%, and 18.26%, respectively [16,18–20] (Figure 7A). All of the above-mentioned restriction endonucleases are members of the PD-(D/E)XK superfamily of Mg^{2+}-dependent nucleases (Figure 7A) [10].

A

```
R.Csp231I       MMKVVKMVKIPPLAQDCDLDSSELSEFTPEEHFNKSITRWFSDHYASYSNRFESYEYIQN 60
R.HindIII       --------------------------------MKK-----------SALEKL---LSLIE 14
R.LlaCI         ------------------------------------------------------VVYNLE 6
R.EcoVIII       ----------------MDL--------TNIEFIND-----------AAINRRRYWTNEIV 25
R.PsaI          -----------------------------MHVITE-----------AAIKRREYWVEEIH 20
R.PanI          -----------------------------MHVITE-----------AAIKRREYWVEEIH 20

R.Csp231I       YIQNEHFNWS-----VAPNTNVIATKFSAHLRTLSLKDFSFLLCHTGYIPEIYKADSSQE 115
R.HindIII       NLTNQEFK-------------QATNSLISFIYKLNRNEVIELVRSIGILPEAIKPSSTQE 61
R.LlaCI         NLVNSEFEKLKGTGLQTVDTEKVTLDFFKMLKKISDKEFINILITSGYIPDLYVADSKEE 66
R.EcoVIII       KLS-GHFV---------NDSSRVEEEIIYEVSKSGSQALLDHLRLCTAIPESYEHDSSEE 75
R.PsaI          KISEGSFG---------NKSERLERELAQEITTQGSEALIDHLRLSGDIPESYGHDTTEE 71
R.PanI          KICEGSFG---------NKSERLERELAQEITTQGSEALIDHLRLSGDIPESYGHDTTEE 71
                 :    *                  .:   : . . : .   :     :*:    .:.:*

R.Csp231I       TLYSKLVEAMVNEWALRMKFTYSLLPTQKSSKEDITISDGE--NIIVADAKSYRLGRSQA 173
R.HindIII       KLFSKAGDIVLAKAFQLLNLNSKPL-EQRGNAGDVIALSKEFNYGLVADAKSFRLSRT-- 118
R.LlaCI         TLFTKLCEALEVDWASRMGFEANAV-TQKSSYEDVVIKINN--KIIVSDTKSFRLGRSQQ 123
R.EcoVIII       KLYSKYTDALISECFKFFGLNSIVL-TERADAADVEVVCDS--YSFVADAKVFRLSRT-- 130
R.PsaI          KQYSKYTDALLSETYKALGMNSIVL-KERADAADVEVVANG--FSFVADAKAFRLSRT-- 126
R.PanI          KQYSKYTDALLSETYKALGMNSIVL-KERADAADVEVVANG--FSFVADAKAFRLSRT-- 126
                . ::*  : :  .    : :    :  ::..  *:          :*:*:* :**.*:

R.Csp231I       APNVKDALKKGDITKWLAYYDQHKYNRIGGLVAFPSQHDWKNGSDFYLYLTDKNSPIIML 233
R.HindIII       AKNQKDFKV-KALSEWRE-------DKDYAVLTAPFFQYPTTKSQIFK--QSLDENVLLF 168
R.LlaCI         APNVKDFVKPEDYSKWAN---RHSGQKLGGLVVYPQLHEWTRKSDAHVYCSDKKNPILML 180
R.EcoVIII       AKNQKDFKV-QAMDGWRN-------TKDFAMVVCPIYQLPVKSSQIYQ--QAILRNVCVF 180
R.PsaI          AKNQKDFKI-QAMDGWKR-------GKPYAMVVCPIYQLPTKSSQIYE--QASSRNVCIF 176
R.PanI          AKNQKDFKI-QAMDGWKR-------GKPYAMVVCPIYQLPTKSSQIYE--QASSRNVCIF 176
                * * **         *          :  .::. *  :     *: .         : ::

R.Csp231I       FYEHMAFMLLAGMDKN-N-----LLDFYRNHKDIFPNEVFNKIGSRKIYFDKLEQYLLNC 287
R.HindIII       SWEHLAILLQLDLEETNI----FPFEQLWNFPKKQS------KKTSV--SDAENNFMRDF 216
R.LlaCI         PYHYLAYFLERKDKFNPK-----SLEKLWDYEKIFPEK----ADSRNDYWQKINNVILEI 231
R.EcoVIII       TYTHLAVLVKYADIVGS-DVRILLEEIFRSVILMNP------SKDSVQYWTMINRTMLNY 233
R.PsaI          TYSHLAMLVSFSQLESPARAEELIHHVFQTVPALNP------SKDASQYWLSINRAMLGF 230
R.PanI          TYSHLAMLVSFSQLESPARAEELIHHVFQTVPALNP------SKDASQYWLAINRTMLGF 230
                 : ::* ::                .                           :. :

R.Csp231I       GKFKWSEFSETSDLI--------------------------------------------- 302
R.HindIII       NKYFMDLFKIDKDTLNQLLQKEINFIEERSLIEKEYWKKQINIIKNFTREEAIEALLKDI 276
R.LlaCI         TGDEKKEFKKFLNLAET----KLYEFV-------------EGRLKNLEYQKNI----KIK 270
R.EcoVIII       DKRIEKLWIDERIATSEG----IYVLK---KMAIEYLSSERGRILSMSREEAVRALIKMN 286
R.PsaI          SDKISELWRIEKQAAVES----IAVAK---EEALMFLAREREKIMRLSHQDAIKELINMH 283
R.PanI          SDKISELWRIEKQAAVES----IAAAK---EEALMFLAREREKIMQLSHQDAIKELINMH 283
                     . :

R.Csp231I       --------VSEKALRHSY---R-------------------------------------- 313
R.HindIII       NMSSKIETID------------------------------------------SFIKGIKS 294
R.LlaCI         KIEFEISSIPDSELRDKFLKYRQEIETQYIVTFQERIQKFRLTNNKESTTYSKFIDSSFD 330
R.EcoVIII       KIESRIEQIK------------------------------------------K----VTD 300
R.PsaI          KIESRMKVIS------------------------------------------Q----VSD 297
R.PanI          KIESRIKVIS------------------------------------------Q----VSD 297
                        :

R.Csp231I       ------- 313
R.HindIII       NDRLYL- 300
R.EcoVIII       NNILSLK 307
R.LlaCI         KS----- 332
R.PsaI          NGLLEIR 304
R.PanI          NGLLEIR 304
```

Figure 7. *Float*.

B

```
M.Csp231I       ----------MRKQLLINADSISELKKLEDNSIDLILSDIPYGIGADDWDVLHKNTNTAY 50
M.HindIII       -----------MIDCIYNSDSICEIKKLDSNSIHAIISDIPYGIDYDDWDILHSNTNSAL 49
M.LlaCI         ----------MKIDTIYNENSIDSIKKIETESIHLILSDIPYGISFADWDILHENTNSAL 50
M.EcoVIII       MLNKIQLPDEIKKNTVTNGNSISLIKKLPSQYAHLILSDIPYGIGAEDWDVLHNNSNNAY 60
M.PsaI          --------MEDLYNKLFVGDSITGIKRLPDNFVDLILSDIPYGIGAEDWDVLHDNTNSAY 52
M.PanI          --------MEDLFNKLFVGDSIAGIKRLPDNFVDLILSDIPYGIGAEDWDVLHDNTNSAY 52
                             : :   :**  :*::  :  . *:*******.  ***:**.*:*.*

M.Csp231I       LGNSPAQKQAGNVFKRRGKPINGWSDADKKIPAEYYQWCQTWASEWLRVLKPGGTAFVFA 110
M.HindIII       GGTSSAQHKT-SLFKRRGKPLNGWSEADKKRPQEYQEWVESWSNEWFRVLKSGSSVFVFA 108
M.LlaCI         GGSSPAQHKTGSGFKMRGKPINGWSEADRKIPLEYQNWVESWAKEWFRVLKPGSSCFIFA 110
M.EcoVIII       LGSSPAQKNAGAIFKKRGKPINGWSEADKKIPLEYQQWCEEWATEWFRVLKPGASAIIFA 120
M.PsaI          LGSSPGQQKAGAIFKKRGKPINGWSEADRAIPRQYYEWCSTWAPDWLRTLKPGGSVFIFA 112
M.PanI          LGSSPGQQKAGAIFKKRGKPINGWSEADRAIPRQYYEWCSTWAPDWLRTLKPGGSVFIFA 112
                 *.* .*:::   ** ****:****:**:  * :* :* . *: :*:*.** *.: ::**

M.Csp231I       GRRFAPRCIVALEDAGFNFRDMLSWIKPKATHRAQRLSIVYERRGMQDESLKWNGWRIGN 170
M.HindIII       GRQFAHRVVVAFENSGFTFKDMLSWEKDKAPHRAQRISCVFERRGDIANTNKWVGWRVAN 168
M.LlaCI         GRRYAHRAIVGLENSGFTFRDMIGWNRSKATLKAQRISKVYERRNDFENAEKLSEWRVGN 170
M.EcoVIII       GRRFSHRCICAMEDAGFNLRDIIAWMRVKAPHRAQRLSCVYERRGDTLNAEKWNGWRVGN 180
M.PsaI          GRRYAHRCIAAMEDAGFSFKDMFSWMRQKAPHRAQRLSIVYQRREDHANAVKWEGWRVGN 172
M.PanI          GRRYAHRCIAAMEDAGFSFKDMFSWMRQKAPHRAQRLSIVYQRREDHANAVKWEGWRVGN 172
                **::: * : .:*::**.::*::.* : **  :***:* *::**    :: *   **:.*

M.Csp231I       LRPIFEPIIWCFKPYKH--TIADNVLEHNLGAYNQFAFEEITG-----HFNNTLEIGMSP 223
M.HindIII       LRPLFEPILWFQKPYKTGSTLADNLIKHEVGAWNENSLTHWNIQQGALNHSNILKVRITS 228
M.LlaCI         LRPVFEPILWFTKPYKQGGTIADNMLKHGVGAYNLEKWQTFSE-----KGDNYIEIPNLS 225
M.EcoVIII       LQPTFEPILWFSKPYKIGGTIADNAIIHGVGAYNQDAFVARNG-----KPENVITAGFSS 235
M.PsaI          MRPTFEPVLWFTKPYKIGTTIADNVLTHGVGAFNEEAFLKYEP-----TPDNVLTCGFEA 227
M.PanI          MRPTFEPVLWFTKPYKIGTTIADNVLTHGVGAFNEEAFLKYEP-----TPDNVLTCGFEA 227
                ::* ***::*  ****   *:*** : * :**:*                .* :

M.Csp231I       KEGGLHDAQKPVKLLETLISLVTIPGQVVLDPFAGSGSTAIAALNLKRDFIMIEKDPNIF 283
M.HindIII       EDKGYHVAQKPLNLMKLLIDLVTKEEQIVLDPFAGSGTTLLAAKELNRHFIGYEKNNGIY 288
M.LlaCI         SDRGLHPTQKPLVLMKALIELTTQENQIVLDPFSGSGTTLVAAKELNRHYLGFEIDKDYY 285
M.EcoVIII       NESGLHPTQKPVALMKTLIELTTQKGQLVIDPFSGSGSTLVAAKDLGRDYIGFEINPTYV 295
M.PsaI          GETGLHPTQKPVRLMQSLIELTTQEGQIVLDPFSGSGSTLVAAQRLGRKYIGFEASADYA 287
M.PanI          GETGLHPTQKPVRLMQSLIELTTQEGQIVLDPFSGSGSTLVAAQRLGRKYIGFEASPDYA 287
                 : * * :***: *:: **.*.*   *:*:***:***:* :**  * *.::  *  .

M.Csp231I       SIMNDRLSKSKQTISLFD--- 301
M.HindIII       NIAVNRLGIEKNNCFYNKEKK 309
M.LlaCI         NTSLNRLNTLL---------- 296
M.EcoVIII       ETSIKRLNK------------ 304
M.PsaI          KIAEARLEIKKNQPSLFDL-- 306
M.PanI          KIAEARLDINKNQRSLFDL-- 306
```

Figure 7. Comparison of the amino acid sequences of restriction endonucleases (**A**) and methyltransferases (**B**) of HindIII-like R-M systems. The amino acids of the putative catalytic magnesium binding motif PD-(D/E)XK are shown in red. The conserved sequence motifs corresponding to the methyltransferase catalytic motif IV (DIPY) and the AdoMet binding motif I (FGG) are marked. Identical amino acids are indicated with an asterisk; two dots represent a highly conservative substitution, and one dot represents a conservative substitution.

The second gene, *psaIM*, encodes a polypeptide of 306 amino acid residues (M.PsaI) with a calculated mass of 34 517 Da containing conserved amino acid sequence motifs typical for the N^6-adenine β-class methyltransferases [13]. The alignment of the entire M.PsaI amino acids sequence with well-studied isospecific methyltransferases of R-M systems, such as M.EcoVIII, M.LlaCI, M.Csp231I, and M.HindIII, revealed significant homology of 64.47%, 57.77%, 57.47%, and 49.34%, respectively [16,18–20]. (Figure 7B). In addition, M.PsaI also shares remarkable sequence identity with methyltransferases from

Marinospirillum sp. (80.27%), *Candidatus Nitrotoga fabula* (79.25%), *Negativicutes bacterium* (78.86%), and *Cylindrospermopsis raciborskii* (77.85%).

Bioinformatic analysis revealed that a large number of R.HindIII homologs are distributed in several bacteria. However, only a few HindIII-type R-M systems have been well-characterized in *H. influenzae* (HindIII), *Citrobacter* sp. (Csp321I), *Lactococcus lactis* (LlaCI), *Bacillus stearothermophilus* (BstZ1II), and *Escherichia coli* (EcoVIII) [16,18–20].

2.8. Analysis of the Regions Flanking the R-M System PsaI

The flanking regions of R-M systems are often characterized by the occurrence of mobile genetic elements involved in recombinational events [21,22]. This is also the case for the PanI R-M system. Analysis of the nucleotide sequence revealed an ORF, located immediately 5′ of the *panIR* gene, that encodes the magnesium chelatase domain-containing protein, whereas downstream of the *panIM* gene an ORF that is related to a mobility-associated protein (recombinase family protein) is placed. Having determined a complete PsaI R-M system, we made an attempt to identify sequences adjacent to the *psaIM* and *psaIR* genes. For this purpose, two sets of primers were designed; in each pair, one primer was targeted at a known region of R-M system PsaI, while the second primer matched the sequence flanking isomeric R-M system PanI (Table S1). Using a primer set complementary to *psaIM* and a recombinase gene adjacent to *panIM*, we obtained a PCR product of the expected size (900 bp) and determined its sequence. An examination of this DNA fragment revealed the presence of an ORF coding for recombinase that is almost identical (98%) to its counterpart in the PanI R-M system. However, when the second set of primers designed for *psaIR* and the magnesium chelatase domain-containing protein gene adjacent to *panIR* were involved, no band of the expected length was detected.

This observation suggests that a homologous DNA segment containing the R-M complex is present at different chromosomal loci in the two *Pseudomonas* genomes. We hypothesized, therefore, that the presence of a recombinase enzyme involved in DNA mobility in the vicinity of the R-M PsaI system may serve to transfer the R-M gene complex within and between bacteria. Indeed, there have been reports indicating that the presence of a gene encoding recombinase near genes of the R-M system can be an indicator of horizontal gene transfer [21,22]. Additionally, the migration of DNA segments containing R-M complexes at different positions in the genome may be facilitated by extensive genomic plasticity and diversity of the genus *Pseudomonas* [7].

It should be noted that both *P. anguilliseptica* strains have almost the same nucleotide sequence within their R-M systems. Moreover, they are geographically unrelated and have been recovered from different environments. Strain KM9 was isolated from the activated sludge of a WWTP in Poland, while the MatS1 strain was identified in a soil sample from Sri Lanka.

3. Materials and Methods

3.1. Site Description and Sampling Protocol

The WWTP in Gdynia Debogorze receives and treats municipal sewage from Gdynia and other small surrounding towns. It serves about 470,000 people and discharges approximately 60,000 m^3 per day to the sea. Activated sludge samples were collected in polypropylene tubes, transported to the laboratory at a temperature of 4 °C, and processed on the day of collection. Samples were collected in April, May, and October 2012.

3.2. Growth Conditions

Samples containing bacteria were inoculated onto MacConkey agar and incubated at 37 °C for 24 h under aerobic conditions. The next day, the specimens were examined for growth and colony morphology and subjected to Gram staining. One isolate producing a restriction enzyme that cleaves DNA yielding a characteristic pattern was selected for identification. A single colony of bacteria was identified as *P. anguilliseptica* using MALDI-TOF MS (MALDI biotyper; Bruker Daltonics, Billerica, MA, USA) according to the manufacturer's instructions. This strain was routinely cultured at an optimum growth temperature of 25–27 °C in tryptic soy agar (TSA). The bacteria was maintained on TSA agar plates at room temperature and for long-term storage kept frozen at −70 °C in LB broth supplemented with 20% glycerol. For R.PsaI purification, a single colony of *P. anguilliseptica* KM9 was grown in LB broth at 30 °C for 24 h, and this culture was used for further inoculation of 1 L of LB. Erlenmeyer flasks were incubated at 30 °C for the next 24 h, and cells were harvested through centrifugation and stored frozen.

3.3. Restriction Endonuclease Activity Assay in Cell Lysate

The occurrence of restriction endonuclease in bacterial strains was tested using the modified lysozyme and Triton X-100 method [8]. Bacterial cells were collected from Petri dishes and transferred into 20 µL of incubation mixture A, containing 20 mM Tris-HCl pH 8.0, 1000 mM NaCl, 12.5 mM EDTA, 10 mM 2-mercaptoethanol (ME), and lysozyme at a concentration of 10 g/L. The sample was incubated for 30 min at room temperature, and then 20 µL of incubation mixture B containing 20 mM Tris-HCl pH 8.0, 2% Triton X-100, and 10 mM ME was added for 60 min at 6 °C. The restriction endonuclease activity was assayed in 20 µL of reaction mixture containing 0.1 µg of λ or pUC18 DNA, 2 µL of the restriction buffer Tango (Fermentas, Lafayette, CO, USA), and 2 µL of bacterial lysate cleared through centrifugation for 1 min (10,000× *g*). The universal Tango buffer was used because it generally yields high activity for most of the restriction enzyme. After incubation for 20 min at 37 °C, the mixture was treated with 20 µL of phenol to stop the reaction and extract the cleavage products, which were subjected to agarose gel electrophoresis. The fragmentation patterns obtained for DNA substrates cleaved with the enzyme present in the cell lysate were analyzed using the REBASE Tools REBsites [4], NEBcutter [23], and REBpredictor [24].

3.4. Purification of the Native Restriction Endonuclease PsaI

All of the purification steps were carried out at 4 °C, unless otherwise stated. The cell paste (10 g) was suspended in 30 mL of buffer P (10 mM potassium phosphate pH 7.0, 20 mM KCl,1 mM EDTA, 10 mM 2-mercaptoethanol (ME), 5% *v*/*v* glycerol) supplemented with 0.1 mM PMSF as a protease inhibitor and disrupted through sonication at 4 °C in 60 × 10 s bursts. The lysate was clarified through centrifugation (14,000 rpm, 30 min) and applied to a 2.5 × 7 cm phosphocellulose P11 column (Whatman, Little Chalfont, Buckinghamshire, UK) equilibrated with buffer P. The column was then washed with 300 mL of buffer P, and the proteins were eluted with a 200 mL KCl gradient (0.02–1.0 M) in the same buffer. Fractions of 3 mL were collected and assayed for PsaI endonuclease activity. The enzyme was eluted between 0.25 and 0.4 M KCl. The active fractions were then dialyzed against buffer H (10 mM potassium phosphate pH 7.0, 200 mM KCl, 10 mM ME, 5% *v*/*v* glycerol) loaded onto a 2.5 × 3 cm hydroxylapatite column (BioRad, Hercules, CA, USA) and eluted with 200 mL of linear gradient of K-phosphate, pH 7.0 (0.01–0.4 M). After that, the fractions with the highest endonucleolytic activity were dialyzed against buffer C (10 mM potassium phosphate pH 7.0, 20 mM KCl, 1 mM EDTA, 10 mM ME, 5% *v*/*v* glycerol), loaded onto a 2.5 × 22 cm CM Sephadex C-50 column (Pharmacia, New

York, NY, USA), and the proteins were eluted with 250 mL of KCl gradient (0.05–0.8 M). The final preparations of the enzyme were dialyzed against a storage buffer containing 10 mM potassium phosphate, pH 7.8, 50 mM KCl, 0.1 mM EDTA, 10 mM ME, and 60% v/v glycerol and stored at −20 °C.

3.5. Molecular Mass Determination

Sodium dodecyl sulfate polyacrylamide gel electrophoresis (SDS-PAGE) was carried out in gels containing 10% acrylamide for the separation gel and 4% acrylamide for the stacking gel. After electrophoresis, the protein bands were visualized through Coomassie brilliant blue R 250 staining. The Mr of the PsaI was calculated using a calibration curve obtained with the following standard proteins: bovine serum albumin (67 kDa), ovalbumin (43 kDa), carbonic anhydrase (30 kDa), and trypsin inhibitor (20.1 kDa).

3.6. DNA Cleavage Activity Assays

PsaI endonuclease activity was assayed in a 20 µL reaction mixture containing 2 µL of the restriction buffer Tango (Thermo Scientific, New York, NY, USA), 150 ng of λ phage DNA, and 2 µL of column fraction (1 h, 37 °C). The DNA was analyzed on 1% agarose gels in 1× TBE buffer, and the completeness of lambda DNA cleavage was evaluated visually. One unit of endonucleolytic activity was defined as the minimal amount of R.PsaI that completely digests 1 µg of phage λ DNA in 1 h at 37 °C.

3.7. Determination of the R.PsaI Recognition Sequence

Through restriction mapping of lambda and pUC18 plasmid DNAs, we were able to determine the recognition sequence of the purified endonuclease.

DNAs were digested into definite fragments using R.PsaI, and a double digestion reaction was also carried out using a second restriction enzyme, which enabled the localization of R.PsaI cleavage sites. The size of the DNA fragments generated through digestion was predicted using the REBpredictor program available on the website (http://tools.neb.com/REBpredictor/index.php) (accessed on 16 June 2014). The positions of putative recognition sequences were matched with the sites mapped through double endonuclease digestion. Then, the expected cleavage fragment was compared with the observed restriction fragments from R.PsaI cleavage of the DNAs.

To confirm the predicted recognition sequence and the cleavage site, R.PsaI-digested pGEM3Zf(+) plasmid DNA was used as a template for DNA run-off sequencing. Purified DNA was subjected to run-off Sanger automated sequencing (Genomed, Warsaw, Poland) using GS5 and GS6 primers (Table S1). Sequencing data were analyzed using Chromas Lite version 2.6.6 software (Technelysium Pty Ltd., South Brisbane, Australia).

3.8. Characterization of the R.PsaI Activity Optima

The temperature dependence of R.PsaI activity was tested by incubating 150 ng of lambda DNA with one unit of the enzyme at different temperatures (30 °C–52 °C) for one hour. R.PsaI activity in buffers supplied with commercial restriction endonucleases from New England Biolabs (NEBuffer 1, 2, 3, and 4) and Thermo Scientific (TANGO, B, G, O, and R) was assessed by digesting 150 ng of lambda DNA with one unit of the purified enzyme for one hour at 37 °C. NaCl's effect was studied using the same protocol in the presence of different NaCl concentrations (0–250 mM). The effects of divalent cations, such as Ca^{2+}, Co^{2+}, Fe^{2+}, Mg^{2+}, Mn^{2+}, Ni^{2+}, and Zn^{2+}, on R.PsaI activity were investigated. Digestions were carried out in 20 µL reaction mixtures containing a buffer free of magnesium ions (50 mM NaCl, 10 mM Tris-HCl, 1 mM DTT, pH 7.9), 0.33 µg of pUC18 plasmid DNA, 10 mM of specific divalent cations, and a 0.5 unit of the enzyme (1 h, 37 °C).

3.9. Determination of the DNA Modification Status of the P. anguilliseptica KM9 Strain

To test the endogenous methyltransferase activity, genomic DNA of *P. anguilliseptica* KM9 was digested with PsaI restriction endonuclease and its isoschizomers, R.HindIII and R.EcoVIII. The reaction mixture had a final volume of 20 μL that contained 1 μg of lambda DNA and 1× appropriate NEB buffer, and it was incubated at 37 °C for one hour. The digestion products were subjected to 0.8% agarose gel electrophoresis.

3.10. Determination of the Complete Nucleotide Sequence of the PsaI R-M System

The procedure for inverse PCR (iPCR) was performed according to Ochman et al. [25]. The iPCR template was prepared through self-ligation of the Sau3AI-digested DNA from *P. anguilliseptica* KM 9. The resulting ligation products were amplified with primer pairs and sequenced. The sequences of the PCR products were assembled to reconstruct the sequence of a region of interest and submitted to the GenBank database (accession number PQ730136). The details of the primers used are shown in Table S1.

3.11. Bioinformatic Tools

DNA and protein sequence similarity searches were performed using Nucleotide BLAST and Protein BLAST respectively [26], BLAST against REBASE [4], or CLUSTALW [27]. Queries for all available R-M systems were obtained from the Restriction Enzyme database (REBASE) website [4].

Supplementary Materials: The following supporting information can be downloaded at https://www.mdpi.com/article/10.3390/ijms26146548/s1.

Author Contributions: Conceptualization, B.F.-B., I.M. and M.S.; methodology, B.F.-B. and M.S.; software, B.F.-B. and I.M.; validation, B.F.-B., I.M. and M.S.; formal analysis, B.F.-B., I.M. and M.S.; investigation, B.F.-B.; resources, B.F.-B.; data curation, B.F.-B. and I.M.; writing—original draft preparation, B.F.-B.; writing—review and editing, B.F.-B., I.M. and M.S.; visualization, B.F.-B. and M.S.; supervision, B.F.-B.; project administration, B.F.-B.; funding acquisition, B.F.-B. All authors have read and agreed to the published version of the manuscript.

Funding: This research received no external funding.

Institutional Review Board Statement: Not applicable.

Informed Consent Statement: Not applicable.

Data Availability Statement: The data presented in this study are available upon request from the corresponding author.

Acknowledgments: We thank Martin Blaszk for help with editing this manuscript.

Conflicts of Interest: The authors declare no conflicts of interest.

References

1. Loenen, W.A.; Dryden, D.T.; Raleigh, E.A.; Wilson, G.G.; Murray, N.E. Highlights of the DNA cutters: A short history of the restriction enzymes. *Nucleic Acids Res.* **2014**, *42*, 3–19. [CrossRef] [PubMed]
2. Pingoud, V.; Wende, W.; Friedhoff, P.; Reuter, M.; Alves, J.; Jeltsch, A.; Mones, L.; Fuxreiter, M.; Pingoud, A. On the divalent metal ion dependence of DNA cleavage by restriction endonucleases of the EcoRI family. *J. Mol. Biol.* **2009**, *393*, 140–160. [CrossRef] [PubMed]
3. Pingoud, A.; Wilson, G.G.; Wende, W. Type II restriction endonucleases—A historical perspective and more. *Nucleic Acid Res.* **2014**, *42*, 7489–7527. [CrossRef]
4. Roberts, R.J.; Vincze, T.; Posfai, J.; Macelis, D. REBASE—A database for DNA restriction and modification: Enzymes, genes and genomes. *Nucleic Acids Res.* **2015**, *43*, D298–D299. [CrossRef]
5. Withey, S.; Cartmell, E.; Avery, L.M.; Stephenson, T. Bacteriophages-potential for application in wastewater treatment processes. *Sci. Total Environ.* **2005**, *339*, 1–18. [CrossRef] [PubMed]

6. Zhang, L.; Shen, Z.; Fang, W.; Gao, G. Composition of bacterial communities in municipal wastewater treatment plant. *Sci. Total Environ.* **2019**, *689*, 1181–1191. [CrossRef]
7. Silby, M.W.; Winstanley, C.; Godfrey, S.A.; Levy, S.B.; Jackson, R.W. *Pseudomonas* genomes: Diverse and adaptable. *FEMS Microbiol. Rev.* **2011**, *35*, 652–680. [CrossRef]
8. Belavin, P.A.; Dedkov, V.S.; Degtyarev, S.K.H. A method for detecting restriction endonucleases in bacterial colonies. *Appl. Biochem. Microbiol.* **1988**, *24*, 121–124.
9. Roberts, R.J.; Belfort, M.; Bestor, T.; Bhagwat, A.S.; Bickle, T.A.; Bitinaite, J.; Blumenthal, R.M.; Degtyarev, S.K.H.; Dryden, D.T.; Dybvig, K.; et al. A nomenclature for restriction enzymes, DNA methyltransferases, homing endonucleases and their genes. *Nucleic Acids Res.* **2003**, *31*, 1805–1812. [CrossRef]
10. Orlowski, J.; Bujnicki, J.M. Structural and evolutionary classification of Type II restriction enzymes based on theoretical and experimental analyses. *Nucleic Acids Res.* **2008**, *36*, 3552–3569. [CrossRef]
11. Cowan, J.A. Role of metal ions in promoting DNA binding and cleavage by restriction endonucleases. In *Restriction Endonucleases*; Springer: Berlin/Heidelberg, Germany, 2004; Volume 14, pp. 339–360. [CrossRef]
12. Vasu, K.; Nagaraja, V. Diverse functions of restriction-modification systems in addition to cellular defense. *Microbiol. Mol. Biol. Rev.* **2013**, *77*, 53–72. [CrossRef] [PubMed]
13. Malone, T.; Blumenthal, R.M.; Cheng, X. Structure-guided analysis reveals nine sequence motifs conserved among DNA amino-methyltransferases, and suggests a catalytic mechanism for these enzymes. *J. Mol. Biol.* **1995**, *253*, 618–632. [CrossRef] [PubMed]
14. Bujnicki, J.M. Sequence permutations in the molecular evolution of DNA methyltransferases. *BMC Evol. Biol.* **2002**, *2*, 3. [CrossRef] [PubMed]
15. Pathirana, S.N.J.; Elvitigala, D.A.S.; Nanayakkara, C.M.; Suravajhala, P.; Rajapakse, S.; Hettiarachchi, G.H.C.M.; Chandrasekharan, N.V. Identification and *in-silico* analysis of a novel restriction enzyme coding gene from *Pseudomonas anguilliseptica*. *Gene Rep.* **2022**, *26*, 101487. [CrossRef]
16. Nwankwo, D.O.; Moran, L.S.; Slatko, B.E.; Waite-Rees, B.P.A.; Dorner, L.F.; Benner, J.S.; Wilson, G.G. Cloning, analysis and expression of the HindIII R-M-encoding genes. *Gene* **1994**, *150*, 75–80. [CrossRef]
17. Franck, C.J.; Berthe, C.M.; Bernadet, J.-F. Identification of *Pseudomonas anguilliseptica* isolated from several fish species in France. *Dis. Aquat. Org.* **1995**, *121*, 51–155. [CrossRef]
18. Madsen, A.; Josephsen, J. Characterization of LlaCI, a new restriction-modification system from *Lactococcus lactis* subsp. cremoris W15. *Biol. Chem.* **1998**, *379*, 443–449. [CrossRef]
19. Mruk, I.; Kaczorowski, T. Genetic organization and molecular analysis of the EcoVIII restriction-modification system of *Escherichia coli* E1585-68 and its comparison with isospecific homolog. *Appl. Environ. Microbiol.* **2003**, *69*, 2638–2650. [CrossRef]
20. Mruk, I.; Kaczorowski, T. A rapid and efficient method for cloning genes of type II restriction-modification systems by use of a killer plasmid. *Appl. Environ. Microbiol.* **2007**, *73*, 4286–4293. [CrossRef]
21. Oliveira, P.H.; Touchon, M.; Rocha, E.P. The interplay of restriction-modification systems with mobile genetic elements and their prokaryotic hosts. *Nucleic Acids Res.* **2014**, *42*, 10618–10631. [CrossRef]
22. Furuta, Y.; Abe, K.; Kobayashi, I. Genome comparison and context analysis reveals putative mobile forms of restriction-modification systems and related rearrangements. *Nucleic Acids Res.* **2010**, *38*, 2428–2443. [CrossRef] [PubMed]
23. Vincze, T.; Posfai, J.; Roberts, R.J. NEBcutter: A program to cleave DNA with restriction enzymes. *Nucleic Acids Res.* **2003**, *31*, 3688–3691. [CrossRef] [PubMed]
24. Gingeras, T.R.; Milazzo, J.P.; Roberts, R.J. A computer assisted method for the determination of restriction enzyme recognition sites. *Nucleic Acids Res.* **1978**, *5*, 4105–4127. [CrossRef]
25. Ochman, H.; Gerber, A.S.; Hartl, D.L. Genetic applications of an inverse polymerase chain reaction. *Genetics* **1988**, *120*, 621–623. [CrossRef] [PubMed]
26. Altschul, S.F.; Gish, W.; Miller, W.; Myers, E.W.; Lipman, D.J. Basic local alignment search tool. *J. Mol. Biol.* **1990**, *215*, 403–410. [CrossRef]
27. Larkin, M.A.; Blackshields, G.; Brown, N.; Chenna, R.; McGettigan, P.A.; McWilliam, H.; Valentin, F.; Wallace, I.M.; Wilm, A.; Lopez, R.J.; et al. Clustal W and Clustal X version 2.0. *Bioinformatics* **2007**, *23*, 2947–2948. [CrossRef]

MDPI AG
Grosspeteranlage 5
4052 Basel
Switzerland
Tel.: +41 61 683 77 34

International Journal of Molecular Sciences Editorial Office
E-mail: ijms@mdpi.com
www.mdpi.com/journal/ijms

www.ingramcontent.com/pod-product-compliance
Lightning Source LLC
Chambersburg PA
CBHW062050220926
51850CB00010B/2726
* 9 7 8 3 7 2 5 8 5 3 4 5 8 *